BRONCO STRONG

A MEMOIR OF THE LAST DEPLOYED **PERSONNEL SERVICES BATTALION**

By

Angela M. Odom, Colonel, U. S. Army Retired

Bronco Strong: A Memoir of the Last Deployed Personnel Services Battalion by Angela M. Odom
angela@angelaodom.com

Cover Art by: Brandon Elrod

ISBN: 978-0-9981086-2-9 (Paperback)

Printed in the United States

(2nd Edition)

ABOUT THE AUTHOR

Angela M. Odom retired from the US Army in 2015 as a decorated colonel. She served in three separate combat operational deployments in her long and meritorious military career.

Odom graduated in 1988 from the University of Southern Mississippi as an ROTC Distinguished Military Graduate with a BS in personnel management. She earned a MS in Human Resources from Central Michigan and a Masters in Strategic Studies from the United States Army War College. She is an accomplished teacher, life coach, professional speaker, and entrepreneur.

Bronco Strong: A Memoir of the Last Deployed Personnel Services Battalion is her first book and a testament to the challenges of daily life for Soldiers during the Iraqi Surge, during which two members of her battalion lost their lives in service to their country.

Odom dedicates herself to uplifting women and military veterans by sharing proven life experiences that empower them to win. Visit her online at www.angelaodom.com.

DEDICATION

This book is dedicated to each member of Task Force Bronco for your innumerable sacrifices in support of Operation Iraqi Freedom 06-08 (October 2006 – January 2008).

From My Whole Heart,

Bronco 6

And

My son, Ryan Matthew Odom for encouraging me to focus, finish, and repeat small steps until this labor of love was completed.

And

In memory of my loving parents,

Hattie (Odom) Jones (August 27, 1943 – June 28, 2011)

Walthall Odom (September 11, 1939 – April 3, 2015)

MAP OF IRAQ

TURKEY
unnamed peak
Mosul
Arbil
As Sulaymaniyah
SYRIA
Kirkuk
IRAN
Samarra
BAGHDAD
Ar Ramadi
Ar Rutbah
JOR
Karbala'
Al Kut
An Najaf
An Nasiriyah
Al Basrah
Umm Qasr
SAUDI ARABIA
KUWAIT
0 50 100 km
0 50 100 mi

TABLE OF CONTENTS

ACKNOWLEDGEMENTS

My former Command Sergeant Major (CSM) Timothy Paul Livengood and I consistently told members of Task Force Bronco, the 15th PSB, to share their stories with anyone who would listen, once they returned to their home stations. This story is written eight years after we held a memorable inactivation ceremony on 15 April 2008 on the storied 1st Cavalry Division's Cooper Field. The ceremony was made complete after General Vincent K. Brooks shared prophetic words of wisdom, which placed the historic event into context, a rousing cavalry charge by the 1st Cavalry Division's Horse Cavalry Detachment, and the Army band played perfectly on key. My memory is fading as I attempt to recall detailed interactions which brings me a mixture of joy, sadness, anger, and pride. This was the most thrilling 21-month period in my life and the most rewarding in my military career.

My battle buddy, CSM Livengood, who was an awesome trainer, mentor, and coach to all of us in Task Force Bronco, gave all he had, all the time. I thank him for his service in and out of uniform. God rest his soul.

God blessed me enormously when He allowed me to be a mom. Ryan, my 11-year-old son, read and edited these chapters, approved the book cover, provided requested and unrequested feedback on the length and content of the chapters, and noted how much time I devoted to writing this book. He is the biggest inspiration for most of my endeavors. Ryan has so many questions about my time in the military, but I do not have an equal amount of patience to engage in his constant military simulations with Nerf guns. I want to answer his questions on behalf of all of the sons and daughters whose parents are not giving them the answers they want.

Thanks to my beautiful niece, Cherrie, and her gentle husband, Keith Tobias, for graciously allowing me to deploy knowing that my then 16-month-old son would be loved and safe under their watchful care. I appreciate my niece for allowing me to disrupt her first year and a half of marriage.

Thanks to my mother Hattie (Odom) Jones, father Walthall Odom, my stepfather Jesse J. Jones, and my youngest brother Robbye J. Odom for keeping me prayed up during this deployment and throughout my military career. May they each rest in peace.

I am grateful to each of my siblings; Shirley A. Williams, Derrick L. Odom, Sabrina L. Walker, and Walthall L. Odom for continuing to be my biggest fans and supporters. They each believe that their little sister can do anything.

Ryan's four sets of Godparents, Anthony and Retired Major Lorie A. Fanning, Retired Lieutenant Colonel Larry Gordon and Retired Lieutenant Colonel Velma Gordon, Darryl and Natasha Louder, and Darryl K. Mason and Colonel Michelle D. Mitchell, who demonstrate their unconditional love for me and my son. During this deployment in particular, they each checked on, prayed for, and supported both Ryan and me throughout 2006, 2007, and 2008.

We simply would not have a professionally outlined, written and edited book without Carrie Woods and Lindsey Petrie. They served as my writing coaches, editors, proof readers, and accountability partners during this year long process.

As I began to develop this storyline, I reached out to former members of Task Force Bronco. I thought it would be a disservice for me to share only my perspective of our collective experiences. Several of them humbly and willingly shared their stories, while others engaged in great dialogue as we walked down memory lane and provided updates on our respective families and careers. I am eternally grateful. I was struck by how many details they each recalled and how much emotion is attached to the memory of the events that occurred on 10 October 2007 and 28 November 2007. I thought it was just me who still feels a sense of survivor's guilt and remorse for our Soldiers' deaths and physical and emotional injuries. Grown men and women wept while sharing their stories. Several of them remarked that they had not talked to anyone about their experiences.

So many people helped to bring this story to fruition.

Task Force Bronco's nine command teams consisted of five personnel services support detachments, two postal companies, as well as a headquarters detachment and a rear services detachment. The greatest strength that our subordinate command teams possessed was their "Commitment to Service" to our supported population and their assigned and attached Soldiers.

A large majority of our junior leaders had deployed previously and most deployed again after the Operation Iraqi Freedom (OIF) 06-08 rotation. They intuitively understood how to empower, influence, and lead their respective units while always exceeding my expectations. They provided outstanding postal and personnel service support (PSS) to their supported populations. The safety of their Soldiers and Airmen, coupled with the dedication to provide a superior customer service experience was always foremost in their minds.

Our four subordinate organic elements based out of Fort Hood, Texas were the Headquarters and Headquarters Detachment, Alpha Detachment, Bravo Detachment, and Charlie Detachment. I did not meet the leaders or Soldiers assigned to Delta Detachment and Echo Detachment due to the timing and circumstances of my arrival; they were located at Fort Riley, Kansas. Prior to deploying to Iraq, I established a great relationship with our rear detachment personnel. They were comprised of a mixture of heroes and strong leaders who deployed previously or who had a skill set or circumstance necessary to keep the personnel support mission at Fort Hood operating. Two personnel detachments and two postal companies were attached to the 15th PSB during the OIF 06-08 rotation. Our organization consisted of Army and Air Force elements representing all three components: Active duty, Army Reserve, and Army National Guard.

When it was all said and done, Task Force Bronco had command and control over 37 subordinate elements, which deployed on separate timelines from different locations during the extended 15 months that the 15th PSB served in Iraq. Although our Battalion was officially authorized only 177 Soldiers, our highest strength rose to 481 troops assigned to 31 different geographically-dispersed locations across the southern two-thirds of Iraq, covering over 113,058 square miles. The total troops assigned or attached to Task Force Bronco over the 15 months in theater was roughly 1,000. The battalion's four wartime functions were to provide postal operations, casualty tracking, R5 (Reception, Replacement, Return to Duty, Rest and Recuperation, and Redeployment), and PSS to over 96,000 Soldiers, Airmen, Marines and Civilians.

I am forever in the debt of my two battalion Executive Officers (XOs) for their herculean efforts in leading a team of talented and enthusiastic junior leaders on our battalion staff, serving as a sounding board and buffer for our nine subordinate company and detachment commanders, executing my intent expertly, and remaining steadfastly loyal to our Soldiers and to me, their commander.

The two battalion XOs time, while assigned to the 15th PSB represented two distinct phases. Colonel (COL) Jon Althoff led the main body into the Iraqi theater of operations in early October 2006, prior to my arrival. COL Timothy Zimmerman, who assumed his duties in June 2007, led the main body out of the Iraqi theater in December 2007 and subsequently led all of us through a successful inactivation process. As the second-in-charge of the battalion, they often spoke and acted on my behalf. They always had my trust and respect. Jon and Tim provided plans and operational oversight of an up to 481-troop PSB responsible for processing over 132,000 personnel actions, providing postal service to over 1.7 million customers, processing over 9.1 million pieces of mail weighing over

64.8 million pounds, accurately processing and tracking over 4,191 casualty actions, and processing and providing personnel accountability for over 480,000 passengers for inter-theater and intra-theater transport on both fixed and rotary-wing aircraft. In comparison to any other modern PSB during our era, our troopers produced more actions and more of everything. Some of this was because we were in the Iraqi theater longer than the previous PSBs. Another reason was that we managed a geographic area that was previously managed by two separate PSBs.

My most serious-minded Detachment Commander, Lieutenant Colonel (LTC) Seth C. Graves accepted every additional task I gave him with grace. He and his Detachment Sergeant, Sergeant Major (SGM) Anthony Williams, were rock solid. They led the 'Rough Riders' of the Headquarters and Headquarters Detachment. Seth managed the movement of our subordinate units in and out of the theater. He ensured we were able to track small 2-3 person teams of junior Soldiers arriving from God-knows-where to serve alongside Task Force Bronco.

The Commander of Alpha Detachment, 15th PSB now Retired Major (MAJ) Rodney Davis and his Detachment Sergeant, now Retired Master Sergeant (MSG) Scottie Jarrett led the 'Gators' as they conducted six mobile personnel services support missions. They flew a rotating team of four Soldiers into remote outposts to conduct essential personnel services (EPS). I chose Rodney to formally notify the 834th Adjutant General Company's Soldiers of the death of their teammate, Staff Sergeant Lillian L. Clamens. This was the hardest task I gave him during this deployment and he graciously carried it out with dignity and enormous respect.

Lieutenant Colonel Steve Morris and his Detachment Sergeant, Retired MSG Zenalia Patrice Moses led the Bravo 'Bulldogs'. Although their detachment was located away from the battalion headquarters, they proved to be incredibly loyal and adept at teambuilding.

MAJ Richard 'Rick' Strong, the Commander of Charlie Detachment, 15th PSB and his Detachment Sergeant, now Retired Sergeant First Class (SFC) David Hall led the 'Cobras'. They had the bulk of Task Force Bronco's casualty mission and by the end of the rotation assumed the postal mission on Camp Liberty. They managed five casualty liaison teams (CLTs), while providing their Soldiers the resources and guidance necessary to conduct the EPS Mission. Rick was the Adjutant General's (AG) Corps Regimental Association's Officer of the Year in 2007. Indicative of his work ethic and sustained outstanding performance, in 2016, he was recognized as a Distinguished Member of the AG Corps.

LTC Dale Jones and his First Sergeant (1SG), now Retired Sergeant Major (SGM) Robert L. Gassaway III, led the 175th AG Postal Company, an active duty unit attached to the 15th PSB from Fort Lewis, Washington. The 834th AG Postal Company was led by now Retired SGM Anthony Smith. I gave operational oversight of the postal company to LTC Morris and MSG Moses who were co-located with the 834th at Camp Adder in Tallil, Iraq, when it became clear that the company commander was not focused on leading the Soldiers entrusted to him. I thank each of these leaders for stepping into their purpose and leading under fire (literally). The two postal companies managed 15 Army post offices and 16 mobile postal missions. The postal platoons experienced the bulk of the mortar and rocket attacks because they were located on remote outposts and Forward Operating Bases (FOBs) supporting the Brigade Combat Teams.

MAJ Lynnette Jones and MAJ Nicole Bires commanded our two reserve component personnel services detachments. MAJ Jones' unit, Detachment 5, 376th Personnel Services Battalion, was headquartered in Long Beach, California. Major Bires' unit, Detachment 4, 847th Personnel Services Battalion, was headquartered in Fort Snelling, Minnesota. I thank them both for their service and passion for leading Soldiers.

Chief Warrant Officer Four (CW4) Ronnie Haynes and CSM Katrina Herzfeld made the job of leading the Rear Detachment back at Fort Hood look surprisingly easy. They took care of Soldiers who did not deploy to Iraq during this rotation as well as Soldiers who returned to Fort Hood prior to the TOA in Iraq on 2 January 2008. They interacted with family members during the morning, noon, and night. They also turned in all of the battalion's property and equipment. They moved the battalion headquarters out of our long time work area into the 1st Cavalry Division headquarters building. They coordinated use of a portion of the basement area as a temporary work environment until the battalion's inactivation ceremony on 15 April 2008.

Our talented and exuberant Warrant Officers led the training effort throughout our deployment. Thanks to CW4 Roberta 'Tina' Cortez, CW3 Mark Fraley, CW4 Ronnie Haynes, CW5 Retired Ronald Mavity, and CW4 Melba P. Mosley for sharing your vast experience and technical knowledge. We simply could not have done this without either of them.

My sorority sisters have been a consistent source of strength for over 30 years. During this deployment, LTC Shilisa Geter coordinated monthly dinners and gatherings for any members of Delta Sigma Theta Sorority, Inc., who were on the Camp Victory Base complex to meet at the DFAC, eat, and fellowship together. We were all assigned to different units and were from different military installations. We represented all components of the U. S. Armed Forces: Air Force, Army, Coast Guard, Marine Corps, and Navy. Being in the presence of positive, resilient women was priceless. We all cheered for and prayed for one another to push through our shared tumultuous deployment experience.

Others who provided their input include, MSG Jeff Baker, former First Lieutenant Angela Montgomery Capetillo, Retired Staff Sergeant Aaron Douglass, CSM Nyeedra Edwards, Sergeant First Class Steve Graves, Retired Sergeant First Class Ireshekia Henry (formerly Hilliard), COL Angelia Coffman Holbrook, former Sergeant Myrtis D. Laws, Retired Sergeant First Class Elemenia L. Malone, former Staff Sergeant Brandie M. McKinley, Retired Sergeant Craig Morrison, Staff Sergeant Kellee Neal, CW2 Nicole Patterson (formerly Berhens), CW2 Yvonne Ralph (formerly Adams), and COL Yolonda R. Summons.

As CSM Livengood and I traveled our battle space visiting our troops, they always amazed us with their understanding of the significance of their contributions to the fight and the importance to the warfighter of performing their mission well. It was a true team effort. I like to use a football team as the analogy for our role. We were like the defensive line, we may not have gotten all the glory, but we were an important part of what the team needed to win. If we were not successful at what we did, then the whole team would suffer.

INTRODUCTION

"This is the day which the Lord hath made; we will rejoice and be glad in it." Psalm 118:24 (KJV)

The mindset my mother raised me to embrace is to be unbowed and unapologetic, while always being grateful, faithful, authentic, and above all, resilient.

My five siblings and I lived below the poverty line. However, we were happy, healthy, and had all of our needs met. I remember laughing a lot, playing board games, and always engaging in some kind of family gathering. My mother and father divorced when I was three years old. They did some sort of hybrid co-parenting thing which allowed my father to always be in our lives. We grew up in Mississippi during a time when everyone in the neighborhood treated you like family. We played kickball, red light/green light, hide and go seek, hopscotch, and all kinds of other playful games at the Grove Park Community Center, which was located across the street from our 3-bedroom, 1-bath house. We were not afraid of anything except stray dogs and the mean old lady who didn't like kids walking on her grass. If we were not at home by the time the humming noise the street lights made when they kicked on filled the air, kids would break out in a full-fledged sprint. We had to be in the house before dark. If our mother ever screamed our first, middle, and last name throughout the neighborhood, we braced ourselves for a not-so-pleasant reminder as to why we should have been in the house. We grew up playing outside with our cousins at our grandparents' homes, which were approximately two miles from one another.

The first and only time I defied my mother was at age 18. I signed my Army enlistment contract in December of 1984 after she refused to sign when I was 17; to enlist before I turned 18, I needed parental consent. When I talked to her about joining the U. S. Army Reserves and earning money so I could stop being a burden to her, she refused to have the conversation. So, I waited until after my birthday in November to meet the age criteria. I told my mother I would be attending basic training in June of 1985 after my first two semesters at the University of Southern Mississippi. I told her one of my high school classmates, Myrtis Laws, and I would be traveling to Fort Jackson, South Carolina together. The consequence for my act of defiance and independence was a good, old fashioned beat down.

A few of my mother's brothers served in World War II and the Korean War. My stepfather, who served in Vietnam, was a humble Purple Heart recipient. My mother and stepfather both retired from the Veterans Administration Hospital in my hometown. Her experience of how Americans viewed and treated women veterans was not very positive and she did not want that for her baby girl.

At the beginning of my sophomore year at the University of Southern Mississippi, I clearly remember my mother telling me that she could not help me pay my college tuition. Along with working the drive-thru at the local Dairy Queen and as a hotel maid, I was fortunate enough to work as a Resident Assistant in Roberts' Hall while I served as the chapter president of my undergraduate chapter of Delta Sigma Theta Sorority, Inc.

In August 1985, I was grateful to earn a three-year Army Reserve Officer Training Corps (ROTC) scholarship to pay my college tuition. Participation in the ROTC program decreased the financial burden on my mother and increased my self-efficacy.

In 1988, I was fortunate to become a United States Army ROTC Distinguished Military Graduate from the University of Southern Mississippi, where I also earned the title of homecoming queen, earned a Bachelor of Science Degree in Personnel Management, and was commissioned as an Adjutant General's (AG) Corps Second Lieutenant (2LT).

I loved attending the University of Southern Mississippi and have many amazing memories from my time there. My fellow students voting me homecoming queen was by far one of the most memorable events. This was the year the National Football League's Hall of Fame quarterback, Brett Favre, was a freshman. I did not have any idea of his remarkable talent and his love for the game. The Ku Klux Klan marched down Main Street in Hattiesburg, Mississippi apparently in protest of this predominantly white university student body electing a black homecoming queen. I couldn't have cared less. I was a college student who was having fun and was eager to start my military adventure in a matter of months.

For my military career, I chose to serve as an AG officer because I understood it to be a multi-faceted field. I knew the officers could serve as commanders and staff officers. I knew the overall mission of an AG officer was to support the warfighter by decreasing their concern about their personnel-related actions, such as their mail, the accuracy of their casualty reports, their promotions, awards, evaluations, and any other administrative actions that were pertinent to morale and to a professional serving in the Army. I knew my parents would feel comfortable with me serving in this career field. Plus, when I initially joined the Army as a private, my recruiter told me that AG admin types always remained in the rear.

That didn't turn out to be true, but at the time it sounded good.

I graduated from the officer's basic course, where I learned the technical portion of my new job as a military personnel officer and a little bit of tactical training. I was assigned to one of the largest and one of the most notable military installations, Fort Bragg, North Carolina, the home of the Airborne Soldier. I was not airborne at the time, however, I was in great physical shape and was eager to learn my job. I was assigned to the 18th Personnel and Administrative Battalion's 19th Replacement Detachment.

As a newly-minted Lieutenant (LT), I did not realize how blessed I was to have my sorority sister and first mentor, now Retired Colonel (COL) Vaseal Montgomery, and my battalion commander (BC), now Retired COL Thomas L. Owens, provide tutelage and guidance during this critical foundational period in my career.

In August 1990, then Lieutenant Colonel (LTC) Owens successfully deployed the 18th Personnel and Administrative Battalion to Saudi Arabia in support of Operation Desert Shield / Desert Storm. This was as a result of the Iraqi dictator, Saddam Hussein, invading Kuwait in early August 1990.

I was a 2LT in the replacement detachment, which was basically a place where everyone who arrived at the installation would initially report prior to being assigned to their work assignments. Then, I was selected to work as COL Owens' personnel officer in the battalion S1 position. Fortunately, I was there for only a few months when Desert Shield started. It seems unreal now, but it literally took months to transport equipment from Fort Bragg, North Carolina to Saudi Arabia.

I was a typical 2LT; trying to learn everything I could as fast as I could. It seemed as if every officer assigned to the Personnel and Administrative Battalion knew my job better than me. At least that's what they told me constantly when they shared their tall tales of being in my position. Looking back, I can say that I probably should have talked less trash and listened more. The BC fired me after being in Saudi Arabia for only a few weeks. I failed to do due diligence and verify the items on the supply list, specifically, we did not pack enough clips that we routinely used to fasten awards to Soldiers' uniforms during award ceremonies. My BC wanted to present an award and I asked where the butterfly clips were and one of my NCOs told me we didn't have any.

In those days, you were allowed to smoke inside of work areas. The BC had a cigarette in his office, which was a repurposed old raggedy white shack in the middle of nowhere, Saudi Arabia. I'll never forget how he took a long pull off his cigarette. When he finished, he balled up his fist and hit the door with enough force to leave it dangling from the frame. He told his second in command, the battalion XO "Get her out of here." I was fired. That was the last straw for him and the absolute best thing for me. My prayers had been answered. The people who worked in the battalion headquarters seemed to always be on edge and stressed. I do not ever recall my immediate supervisor providing any guidance on how to succeed. In retrospect, I was most likely too arrogant to listen if he had tried.

The postal company commander, Retired Major (MAJ) Gerald Hughes asked the battalion commander if I could serve as his company XO. Thank God for his grace. I loved serving in the 129th AG Company. The Soldiers worked long, laborious hours, pitching and processing tons and tons of mail. They took great pride in caring for the warfighters' personal mail. The BC came over often to visit. He would sit on a stump and talk with me and the other lieutenants. He freely gave us guidance on how to work through various personnel issues and how to develop better processes and procedures. He mentored me even after firing me. He also hosted my promotion to First Lieutenant (1LT). I later found out he was also a graduate of the University of Southern Mississippi. This is a fact that was surely something a good personnel officer should have known earlier.

I watched how COL Owens protected his lieutenants and captains (CPTs) from anyone outside of the battalion. He protected our time. I watched as he verbally pummeled his staff when they were not providing resources to his subordinate units. I watched him when he provided updates on his vast PSS mission. I was very aware of when he traveled to the various locations to visit and protect his subordinates who were not co-located with the battalion headquarters. I placed all of these lessons, good and bad, within my imaginary 'kit bag.' My battalion commander was my first military example of a competent, confident, and caring leader.

After returning home from Operation Desert Shield/Desert Storm in April 1991, I was reassigned to the 1st Corps Support Command (COSCOM) on the other side of Fort Bragg, as a part of the overall officer talent management process. The COSCOM G1 (personnel), LTC Joseph Rhett, supported my request to attend airborne school at Fort Benning, Georgia. After successfully completing airborne school, I returned to Fort Bragg and was informed that I would serve on the 1st COSCOM Commanding General's Assault Command Post. I was voluntold; which is a combination of being told you will complete a task and kind of sort of simultaneously volunteering for said task.

I did not realize it was an airborne position, where I still maintained my responsibilities with the COSCOM G1 directorate while participating in a variety of military training operations. One of the most memorable activities was the Prop Blast initiation, where we conducted lots of physical fitness team building and esprit de' corps activities for what seemed like a gazillion hours. Now, in retrospect, it was only a better part of a day or two. Another memorable event was the airborne operations when we simulated an alert for a real world contingency operation. We were notified to arrive at a certain location with our military gear packed as part of the routine requirement to be ready to deploy anywhere in the world within 72 hours. Well, to my ultimate surprise, we conducted airborne operations and landed at Fort Bliss, Texas. I am not actually certain how long we were there. I only remember it being as hot as my grandmother repeatedly told me hell would be. The most rewarding military operation I was blessed to participate in was in support of Hurricane Andrew in Florida. My participation in each of these events was minuscule in comparison to the overall effort that was accomplished.

One of my earliest adult life lessons was taught to me by my mother during one of my visits to Mississippi. I had been in the U. S.

Army for about three years. I worked at Fort Bragg, North Carolina and had driven to Mississippi for about four days to spend time with my mom. I was bragging to my mother about how I had dropped a Soldier, which means I had stopped him and directed him to drop to the ground and start doing the Army's most popular physical training exercise; the push-ups. As I relayed the story to my mother, I noticed she was not smiling. She began to ask me if that was what I was learning in the Army. She wanted to know whether I was being taught to demean people. She asked a series of rapid fire questions; who did I help, who did I support, who did I mentor, who did I uplift? I felt about 2 inches tall after she finally stopped berating me. It was the best counseling I had ever received. My mother taught me to change my perspective and to use my influence for good. Her demeanor took me back to my days as a Private (PVT) in basic training at Fort Jackson, South Carolina.

After my assignment at Fort Bragg, I was reassigned to Fort Benjamin Harrison, located in Indianapolis, Indiana, to attend the AG Officer's Advanced Course. The Army rotates its personnel every 1 to 4 years. The rotation timeline depends on the individual's occupational specialty, practical experience needed for the individual, and the overall mission the Army is required to meet. During this assignment, I was a student. I learned about military personnel doctrine and how personnel actions were processed during wartime operations and during operational assignments. My mother flew to Indianapolis to participate in my promotion ceremony to the rank of captain. I was originally assigned to the Great Lakes Recruiting Battalion, however, after talking with my mother, one of my classmates talked with our assignment officer and swapped our assignments. I was subsequently assigned to the Montgomery Recruiting Battalion, located in Montgomery, Alabama, which is a lot closer to Mississippi than Michigan.

In 1993, I served as the Administrative and Logistics Officer in Charge (OIC) of Montgomery, Alabama's Recruiting Battalion. I frequently went from a hero to a zero in a matter of minutes. This was when Michael Jordan, the hall of fame professional basketball player led his team, the Chicago Bulls to win back-to-back-to-back NBA championships and, at the time, my recruiting battalion held back-to-back-to-back #1 rankings in recruiting civilians to volunteer to join the U. S. Army.

The BC, now retired LTC Jamiel S. Saliba, was an outgoing, happy, very motivational leader. I learned from him that whatever your Soldiers required for their mission, you provided. If you want them to excel, then you don't allow distractions. He was never rude or demeaning, just very positive and encouraging, but stern at the same time. He's one of those guys you don't want to let down. We all felt like heroes working for him and with him. Our higher level of military command provided competitive incentives to maintain excellence. They sponsored Personnel Section of the Quarter and Personnel Section of the Year plaques. Our personnel section earned the plaques consistently because our BC expected our best effort. However, the LTC who replaced him did not demonstrate a will to win.

Our Battalion went from #1 on the list to the bottom of the list of all recruiting battalions in the nation. He wasn't Jamiel Saliba. He allowed his Command Sergeant Major (CSM) to demean, humiliate, and threaten people. I despised working in the unit, however, I knew if I was not there, that the CSM would attempt to wreak holy hell on my Non-Commissioned Officers (NCOs). One of my additional duties was to serve as the urinalysis officer in Alabama and the Florida panhandle. One of the saddest experiences came from learning that one of the recruiters committed suicide right before Thanksgiving. She always appeared to be a happy-go-lucky person. She told me three jokes and we sung an upbeat song together as we walked to the bathroom. I was rushing because when I was done I could drive to Mississippi and enjoy a festive holiday season with my family. We traded pleasantries and wished one another a happy Thanksgiving. By the time I arrived to my mother's home in Mississippi, I received a phone call and was informed that the NCO had taken her own life. I replayed every interaction with her and began to play the 'what if' game to no avail.

During this tour I learned that leadership, positive or negative, makes the difference in whether the mission is accomplished.

My next assignment took me to Fort Wainwright, Alaska to an absolutely fabulous opportunity to command Detachment 1, 203d Personnel Services Battalion for 24 months. I was in my element. We trained in everything from 55-below-zero weather to 60 degrees above. We participated in physical training (PT) exercises outside under extreme conditions, where we would have frozen snot and icicles on our eyelashes.

After relinquishing a successful detachment command, I was reassigned to the Arctic Warriors Infantry Brigade for eight months. A few weeks into my time with the brigade, I thought I had died and gone straight to hell, as I experienced the soft bigotry of low expectations, a lack of effective communication, a blatant disrespect for human dignity, and that was just the beginning. I found out that the culture was for the officers to compete, as opposed to collaborate. As far as I recall, I was the only female of any rank and the only black officer on the brigade staff of about 200 people. A company typically has 100 to 200 Soldiers, and an infantry battalion is a combat unit of 500 to 800 Soldiers. Three to five battalions, approximately 1,500 to 4,000 Soldiers, comprise a brigade. If there were any other black or female officers, I do not recall ever seeing them or having a conversation with them.

I found out that the culture for the NCOs was quite different; they were encouraged to work as a team. Their main concern was taking care of Soldiers and the mission. Your race or gender was irrelevant. They only expected you to carry your own weight, so to speak. This was when I learned that "NCOs are the backbone of the Army" is more than just an Army recruiting slogan.

During this time, my boss, the brigade personnel S1 officer, S1 being a specific group of Soldiers responsible for administrative processing, offered me up to lead a hodgepodge team of mostly engineer Soldiers from different units to construct a dismounted and combined arms, live-fire range in the middle of freaking nowhere. We were flown out to the location and dropped off with 72 hours' worth of meals ready to eat (MRE)s and our personal gear. Thank goodness, I went to the brigade S1 senior enlisted leader (NCOIC) and asked him to check my equipment prior to departing on this mission. The standards were totally different from my PSS unit. We were in an austere environment without logistical support. We slept in 10-man arctic military tents with potbelly stoves that we only ran at night. The potbelly stoves ran without the use of electrical power and could burn many types of liquid fuel. I remember we had to turn the stoves off when we went to sleep because the tents were so dry.

Despite the advantages inherent with the use of heating devices, unique hazards are also presented to Soldiers, especially concerning fire and carbon monoxide poisoning. Fire can engulf a tent in 10 seconds and destroy it in 60, giving Soldiers very little time to react. Conversely, carbon monoxide is odorless, colorless, and tasteless, and can quickly kill Soldiers while they are sleeping. At that time there were recent incidents throughout the Army where Soldiers died in their sleep due to carbon monoxide poisoning or they were burned alive because the fire would burn quicker than they could get out of their sleeping bags. I learned to trust the equipment, my training, and the NCOs.

The only thing I really did not like was pooping in the snow in 55-degrees-below-zero weather. Quite frankly, in my entire military career, I could mentally move past anything except this, but we made it safely through this mission. The brigade commander, I was told, was pleased with the outcome. When I returned to brigade headquarters, I learned that the officers were betting that I would get frostbite or injured in some way.

They lost that bet.

During my assignment in Alaska, I learned to listen more than I talked. The more I listened and observed, the more I realized that the officers assigned and attached to the infantry brigade headquarters did not dislike me, I simply did not matter to them. This was a huge lesson for me because, they did not value diversity; diversity of thought, diversity of race or gender, diversity of Army military career fields. They seemed to delight in the fantasy that the way they viewed the world was the only way the world should be viewed. That's unfortunate.

My final assignment at the rank of Captain was to the Army Central Command (ARCENT), located at Fort McPherson, Georgia, where I served in the both the ARCENT G6 (Communications) and in the ARCENT G1 directorates. This was an awesome mix of diverse professionals of various ranks, genders, and ethnicity. We routinely deployed throughout the Middle East for a few months at a time. The only thing that mattered in this unit was that you provided value and accomplished your specific mission. At the end of this tour, my mother was able to participate in my promotion ceremony to the rank of Major.

Fortunately, I was selected for another institutional training opportunity. I went to the Command and General Staff College located at Fort Leavenworth, Kansas for 10 months. This time, all of the students held the rank of Major and represented different military career fields. On 11 September 2001, we were doing our normal routine. The news was on a projector screen on the wall. We were early for class, so we were sitting there chatting, when we looked up and said, "What is that? What just happened?" The news was saying that a third plane had just hit the pentagon and by that time the hallways were filled with people who were leaving class and calling family members to make sure they were safe because we didn't really know what was going on yet.

After graduation I was assigned to Fort Carson, Colorado, the most pristine place I've ever been. My job was to manage the officer personnel strength of a division-sized element. After about a year of serving in this position, I deployed to the U. S. Central Command forward headquarters in Qatar. I did not deploy as a part of a unit. It was an eye opening experience to deploy to the Persian Gulf as an individual augmentee.

My deployment coincided with the Department of Defense's military campaign titled, "Operation Iraqi Freedom" or OIF for short. A rotation of cohort units would remain in Iraq for about 12 months. I believe our strategic leaders did not anticipate the military conflict to drag on for as many years as it has based on the perceived success of the 100 hours war during Operation Desert Shield / Desert Storm in the early nineties. Each rotation of cohort units were designated as OIF I and II (as in the second rotation of cohort units). During OIF I, I was deployed to the Qatar from March until May 2003. I recall watching a news channel and hearing the President of the United States incorrectly declare victory. He was standing aboard the aircraft carrier USS Abraham Lincoln off of the California coast. We redeployed from Qatar to MacDill Air Force Base, located in Tampa, Florida. He obviously spoke too soon.

I was released from CENTCOM in December 2003 and was assigned to the U. S. Army Human Resources Command (HRC). At the time HRC was located in Alexandria, Virginia. During this assignment, I was blessed to become a mom, was blessed to have my mother participate in my Lieutenant Colonel Promotion ceremony, and was blessed to be selected to serve as a battalion commander (BC).

And that is where the story of Task Force Bronco really begins.

CHAPTER 1

The Journey of 7,000 Miles

"The only thing that is constant is change."

- Heraclitus

This wasn't supposed to happen. I mean, as a Soldier I knew the risks, and as a BC I likely knew them better than most. Yet it still shocks me to think that after 12 months of being in a combat zone and focusing everything I had on supporting our troops, suddenly…I was the one in need of help. Because this wasn't supposed to happen.

I'm getting ahead of myself, but truthfully it's hard not to. When I think about that time in my life and everything that was swirling out of control around me, it is hard to make the pieces fit. To this day it is difficult for me to think about what happened and fully understand.

During the time we spend together in these pages, the words may resonate deeply or only casually; depending on your personal and professional experiences and my writing style. This is a story that deserves to be shared. More so, these are the heroes who deserve to be recognized and remembered.

So let's back up a bit, because to fully comprehend the impact of how this story ends, you have to know how it all began.

On 10 October 2007, my battalion of personnel support Soldiers were pulled from their roles behind the scenes and thrust front and center, into a nightmare. In the blink of an eye, amongst one of thousands of mortar attacks that had taken place during the Global War on Terrorism, my Soldiers became the ones in need of support.

What was difficult at the time for me, and is still difficult when I think back on it now, was hearing our Commander-in-Chief touting success. He and the civilian leadership claimed the overall number of deaths among coalition forces had decreased significantly, but regardless of their words, my Soldiers were still gone.

There was nothing I could do about it and likely nothing I could have done to prevent it, but the thought of "*what if*" is a haunting path to walk. It keeps me up at night....still.

It's a fact that an overwhelming majority of Americans killed or physically wounded in the major conflicts since the Revolutionary War reside among occupational specialties of Infantry, Armor, Artillery, and Aviation. This is what we learn about in books and these are the stories we watch at the movie theater, or on YouTube and Netflix, the tales of those men whose military occupational specialty is thought of as being "*outside the wire.*" They are usually portrayed as valiant warriors who are kicking in doors, diving on a grenade to save their buddies, or shooting their way out of a heavily contested firefight. These heroes are glorified because of their contribution to major combat operations.

This isn't the story I'm sharing with you today.

Today I want to share the story of *my* Soldiers. I want *you* to know that these lives mattered and they still matter to so many of us who hurt for their loss each and every day. I can promise that the number of copies sold of this text is irrelevant. It does not matter to me in the slightest. What matters is that my Soldiers' collective stories are shared.

I want to describe these events from their vantage point and to share what those days were really like. I want to answer some of the questions I know the children and family members of my Soldiers are asking about their loved one's time in Iraq. I want you to know that being given the opportunity to share this story is an honor; one I do not take lightly.

And, in doing so, we will acknowledge the contributions of Staff Sergeant (SSG) Lillian L. Clamens, Sergeant First Class (SFC) Ireshekia Hilliard (Now Henry), U. S. Army Retired CSM Timothy P. Livengood, and SFC John J. Tobiason.

So let's go back, all the way back to the beginning, before I was even a part of their story.

Near the end of the year 2004, senior officers reviewed my previous job performance and military personnel files and selected me along with several other officers in my rank and pay grade to serve as battalion commanders. Because the Army was in the process of taking personnel battalions out of its inventory, there were only a few of these battalions available. This was the only category I competed in. I was selected and my name was placed on an alternate's list. I would be used if the officers who were selected first, were for whatever reason, unable to perform. It was a long shot.

In March of 2005, the 15th Personnel Services Battalion (PSB) left Baghdad, Iraq on a roughly 7,000-plus mile trek to make their way home to Fort Hood, which is located in Killeen, Texas. These Soldiers were tired, they were weary from standing strong through a long 12-month deployment to a war zone, and they were more than ready to see their families, friends, and loved ones.

I can imagine the scene because, even though I wasn't there with this particular unit at the time, I have witnessed and been a part of many such reunions on previous combat deployments. As the planes land at West Fort Hood's Robert Gray Army Airfield you can feel the excitement growing in the air, like electricity, almost sparkling as the families gathered to watch the giant aircraft taxi to a stop.

Then, the doors open and the Soldiers disembark from the plane. Anticipation is buzzing through their veins as they board a long line of buses for this final and shortest leg of their journey home. Together with their fellow Soldiers they will travel to the public welcome home ceremony at Fort Hood on the 1st Cavalry Division's storied Cooper Field.

They exit the buses, turn in their weapons, and plaster big ole' smiles on their faces. The exhilaration is heartfelt. It's a giant relief to be home, to be away from the constant threat of danger and to finally feel completely safe. Now they are only moments away from seeing their wives, husbands, kids, and parents, and wrapping them in their arms, holding tightly as they try to make up for the months of being gone.

They've missed so much, given so much, and they don't want to waste a single second. I can promise you it's a beautiful moment. One that you carry for a lifetime and can rarely recall with a dry eye.

This unit came home under the leadership of LTC Colin Hood and CSM Timothy P. Livengood. These men took their troops to Baghdad, accomplished their Personnel Service Support (PSS) mission in grand fashion, and brought them home safely; the true goal of any command. Now these men had the honor of watching their troops march on Cooper Field in front of a stunning, picturesque background. All four companies marched forward with their company commanders, each moving in sync with their First Sergeants and guidon bearers.

Even though I can sit here and picture it, I truly wish I could have been a part of this moment. It was beautiful. I'm sure of it. And I would love to carry that kind of joy with me, especially as we move through the rest of the story.

What these troops didn't know on this beautiful March afternoon is that they would soon deploy again. And not simply given a new mission, but quite literally sent back to the exact same reality they had just left.

In October of 2006, the 15th PSB would redeploy back to Baghdad with the cohort units supporting OIF 06-08 (Operation Iraqi Freedom 2006 through 2008). This tour would end up lasting a grueling 15 months, with these Soldiers conducting a Transfer of Authority (TOA) ceremony on 2 January 2008. For many of them this new deployment must have felt like an unending reel of déjà vu because they were back in every sense of the word. Back to the same Forward Operating Base (FOB), back to the same Pad 11 at Victory Base Complex in Camp Liberty, back to the same buildings, the same jobs, and even the same bunks.

The only real difference between this tour and the first....was me.

In August 2006, my boss told me to contact the officer who was in charge of managing all of the assignment opportunities for lieutenant colonels and below in my career field. His title was, "the Adjutant General Corps' Officer Assignments Branch Chief." I know, it's a mouth full. I worked in the next building over from his in Alexandria, Virginia. My job at the time was to update Army senior leaders in the Pentagon on the overall enlisted personnel strength. The branch chief informed me that a Personnel Services Battalion was available for command. He further informed me that the unit was in the process of deploying and that I would have to accept the command and prepare to deploy to Iraq within 60 days.

I did not say yes immediately. How could I? There were so many questions running through my head in that moment that it became difficult to have a coherent thought, much less make a spur-of-the-moment, life-changing decision. I told him I would let him know.

He told me I had 24 hours.

I returned to my office, closed the door, and slid down the wall sobbing. I knew instantly that I could do it; I could lead AG Soldiers during combat operations. I knew I could build teams and communicate with our chain of commands. My only fear was leaving my 14-month-old son. This precious baby boy was the center of my life and so newly given to my care. The thought of leaving him behind felt as if my heart was being ripped from my chest.

But as a Soldier, this is what I signed up for. It wasn't about me; it was about the constitution, it was about standing up for my country and answering the call of duty again.

My first phone call was to my niece; a beautiful newlywed who had exchanged her vows less than 30 days earlier on 9 July 2006. She had recently told me, out of the blue, that she would take care of my son if anything ever happened to me. My hand shook as I dialed the number and through my tears I asked her if she was still willing to take care of my son. With no hesitation, she said yes. I told her to stop and ask her husband. I was adamant that I wanted my son in a home where he was wanted and welcomed. I needed him to be loved and supported no matter what.

Again, there was no hesitation.

Their unwavering support gave me the strength I needed to say yes. This was at a time when female Soldiers were in the headlines for refusing to deploy and refusing to leave their children. As a mother, I understand that internal battle. You are absolutely torn between two crucial roles in your life, and without family stepping forward to love my son unconditionally, my decision would have been much, much harder.

If not impossible.

The next call was to my Assignments Manager. I informed him that I would accept the command opportunity. Within 48 hours I received a call from Major General (MG) Joseph F. Fil, Jr., the Commander of the 1st Cavalry Division, located at Fort Hood. He welcomed me to the division and informed me that he needed me there within the next 30 days. I respectfully told him that I would need more time to get my son settled and to coordinate the multi-state relocation properly. He agreed.

My niece and her husband moved from a two-bedroom apartment to a three-bedroom apartment to accommodate my son. We took care of the guardianship documents.

And I left.

In October 2006, I was a 39-year-old doting mother of a 16-month-old healthy, toddler. I arrived in Texas on the Friday of the Columbus Day holiday weekend. My new battalion was in the process of deploying, meaning that my Soldiers had already conducted intensive training to refresh and update tactical and technical skills. They had already conducted "lanes" training where they simulated battlefield conditions, as well as simulated rollover vehicle training along with the myriad of mandatory training every deploying Soldier was expected to know prior to moving into the Iraqi theater of operations. They had already received dental exams, routine medical evaluations, and immunization updates intended to maintain their personal and unit readiness level at all times. They were healthy, physically prepared, and had bonded with one another; they were ready to deploy back to a combat zone together. I, on the other hand, was preparing to lead them on this deployment without even running one training exercise with them. I knew no one, but I was confident that I would get to know everyone quickly; I knew I was ready to assume command of the 15th PSB.

It was time for another ceremony. Much like the one we talked about earlier, but this time in reverse. Now the tears that fell were not tears of joy, but of a profound sadness that has a tendency to stick with you throughout the duration of the deployment. Soldiers who felt like they had just returned to their daily lives were again leaving everything they loved. I felt it was my responsibility to keep my Soldiers safe, to make sure they came home to their families and came back to the everyday lives they loved when they were not serving their country on foreign soil. As I assumed Command of this proven battalion, I felt the weight of my responsibilities and I was ready.

Within the first days of my arrival at Fort Hood, I found out that the previous BC had been relieved of his duties. The Division Commander had lost faith and confidence in his ability to lead and decided that he would not be leading the Battalion back to Baghdad during this deployment. At this time, the advance party had already deployed to Baghdad and the main body was in the process of leaving.

I deployed from Fort Hood to Kuwait initially to in-process into the Theater Gateway and continued to forward deploy into Iraq. The XO, MAJ Jon Althoff, and CSM Livengood led our brave men and women until I arrived. While I was a new face to the battalion this time around, most of the other faces were all too familiar with these surroundings.

Arriving at the battalion headquarters is like traveling to a different world. Our home base was located at Camp Liberty, the FOB in Baghdad near the airport. Everywhere you looked you saw an endless, sandy sea of light brown, made up of brown gravel roads, dirty-looking brown buildings, and rows and rows of trailers, dirty civilian and tactical military vehicles, concrete bunkers, concrete t-walls, and a constant stream of Soldiers all dressed in the same desert camouflaged shade.

What most people don't realize is that this type of established FOB becomes its own small city. We weren't camping in tents stuck out in the middle of a sand dune, instead we had many of the same amenities we would enjoy if we were on base back in the states. The FOB had hot water, showers, a Dining Facility (DFAC), a chapel with worship services, a retail store called a Post Exchange (PX), small medical clinics, post offices (which we ran), gyms, and a mixture of gravel and paved roads.

For my returning Soldiers, the base was the same; the situation however, was not.

Between Operation Iraqi Freedom (OIF) II and OIF 06 – 08, the environment on the FOB had changed dramatically. This, simply speaking, was war.

As soon as we arrived we found ourselves in a combat environment with mortars being fired at random times into every area of our base. The post office, the living quarters, the PX, the DFAC, nowhere was completely safe.

We were like sitting ducks. And I feel that it is important for you, as the reader who may or may not have any similar real world experience, to take a moment and consider what this was like. Imagine being at work, or laying in your bed, or even sitting wherever you are right now reading this book, and knowing that at any moment the world could literally explode around you.

Without warning.

One minute you are going about your day, checking things off your to-do list and then the next thing you know you are huddled in a concrete bunker with your ears ringing. It sounds extreme, because it was. Our reality was we never knew what the next day would hold. All we knew was that our Soldiers were depending on us to provide a positive command climate and ensure that they had all of the required resources to get their job done. They needed our support if they were going to be effective in their specific area of operations. We were determined to give them everything they needed to be successful.

So we worked. We kept our minds focused on the goal and tuned out the noise so we could keep working on our prescribed mission sets. Because no matter how chaotic things became, no matter what, our Soldiers deserved our best.

And they would get it.

In response to the escalated situation and with the use of Improvised Explosive Devices (IEDs) by our enemies on the rise, ground travel was extremely dangerous. So, I demanded that none of my Soldiers ride in convoys. The Brigade Combat Teams' vehicles were being retrofitted with additional armor. My tiny personnel services battalion was not anywhere near the top of the list for military vehicle upgrades. Additionally, the motor Sergeant was doing his best to get the replacement parts we needed to keep our small fleet operational. The risk was simply not worth it, especially with air assets readily available to us.

So, at my insistence, my Soldiers flew from place to place just like the CSM and me. During our time together CSM Livengood shared with me that in the previous deployment IED incidents were not a significant concern. So when he and his Soldiers needed to travel, they participated in road convoys and traveled throughout the country in military vehicles. Hearing him talk about this practice always seemed so surreal to me because it was so vastly different from our current deployment.

For civilians back in the States, travel is not something you worry about too much. You get in your car, you drive to work and get out, and then after work you get back in and drive home. In a combat zone, travel is one of the major concerns, and because Iraq is a country with roughly the same land area as California, travel is often necessary, which is why I wanted to be as safe as possible. And I wanted the same level of safety for my Soldiers.

Because of the size of our physical area of responsibility in Iraq, my CSM and I traveled throughout the country constantly. We needed to make sure the leadership on the FOBs our Soldiers were assigned to throughout the country knew our Soldiers had their own chain of command. We also needed to ensure that the units we supported felt they were receiving our best efforts and that our Soldiers were seeing results from their hard work.

While visiting each location, we interacted with our Soldiers by pitching mail with them and having them demonstrate how to process an award or evaluation, as well as make an identification card for a customer. We asked them how they were being treated and if they needed anything. We listened to them talk about their families and we chatted with them about ours. We ate meals with them and participated in their re-enlistment and promotion ceremonies. We directed our battalion staff to set up video teleconferences for their ceremonies so their families could witness their big moments. We consistently encouraged them throughout the entire deployment.

The point of all this travel was so our Soldiers knew that we cared and that we were there for them. I didn't want to be this obscure name that they only saw during ceremonies or heard of through the grapevine. I wanted to be a person, their leader that they knew and could reach out to if there was a need. More so, I wanted them to feel like they were a part of a team. Combat is stressful and can be extremely isolating and lonely. Many of my Soldiers were very young and all were away from their families. And, while I could not, and would not want to replace their loved ones at home. I did want them to feel like they had a sort of family surrounding them during our deployment. I never wanted them to feel alone or abandoned.

So we traveled, and we talked, and we worked, *together*. Each and every day.

And believe it or not, some days were fairly standard. There is a "typical day" even in this type of atypical environment.

A typical day in Iraq consisted of our Soldiers performing personnel accountability and verifying all sensitive items, such as their weapons, followed immediately by participating in some form of physical training (PT) at or about 5am at their respective job locations throughout the southern two-thirds of Iraq.

We ran Army Post Offices similar to the type of stand-alone Post Offices that you would find in the States. This meant collecting, distributing, and sending the mail, selling money orders and stamps, and all of the other logistics attached to ensuring our Soldiers received their care packages and letters from home and were able to send their messages back to their families.

Other units within our team processed casualty reports and performed duties as casualty liaison officers at the various level-two and level-three medical facilities throughout our area of operations. The point was to track any casualty from the unit level to Germany or back to the states.

Additionally, they performed essential personnel services (EPS) for units. Although Soldiers were deployed, their personnel files were still being reviewed for promotion boards, and evaluations and awards were still being processed. PT tests were still required. Transfers and discharges needed to be accounted for. Leave had to be tracked, approved, and filed. ID cards got lost and were replaced. Soldiers had to be paid and errors corrected. Life in the States doesn't stop just because you are deployed, and deadlines still mattered. All our functions were to support the warfighters. We intended to support them so that they could focus on completing their missions instead of being distracted by personal details. Distractions in that environment are deadly.

Our job, at its heart, was to take care of the details. To handle administrative actions, count the number of troops moving in and out of the theater, track any casualties, and process their mail. And we were good at it. We were dedicated to supporting our brothers and sisters-in-arms so they could keep their focus where it needed to be.

Our goal was to do our job well and then to go home. But, as I've already shared with you, things didn't turn out exactly as we had planned.

CHAPTER 2

A Day in the Life

"Change is hard, and the normal human reaction is to deny it. But, if change scares you, you need to get out of here."

- Lieutenant General Frederick C. Vollrath, Deputy Chief of Staff for Personnel (April 1997)

When you arrive in a war zone you start to develop a new "normal" to your daily routine. Believe it or not, even in that type of high-intensity, high-stress environment, the days begin to blend together until it becomes difficult to recall what you were doing at a specific date and time. And yet, there are still those moments that seem as though they happened only yesterday. Those times you can remember with perfect clarity even years later, and all too often recall during the dark, lonely hours of the night.

For me, one such moment came just a few short weeks after I arrived in the Iraqi theater. I vividly remember uncasing the battalion's colors during the TOA ceremony with the 502d PSB on 15 November 2006 in front of the battalion headquarters on Camp Liberty's Pad 11.

When the announcer called out the portion of the script that outlined our upcoming mission I remember thinking, "Wow, we are going to be extremely busy!" The 15th PSB assumed the mission in support of the Soldiers of the Multi-National Corps-Iraq (MNC-I)

Multi-National Division-Baghdad (MND-B), Multi-National Division-Center South (MND-CS), Multi-National Division-Southeast (MND-SE), and Multi-National Force-West (MNF-W), a full alphabet of 26 locations which all fell under my direct responsibility.

As the words kept coming, my mind was literally racing. It didn't take long for me to start getting anxious because the more I thought about the mission at hand, the more I felt the need for this ceremony to end quickly. I had a job to do and the very first step was for this ceremony to end so I could get off the Pad as soon as possible and start getting to know my Soldiers.

At that time I wasn't sure how effective we were in performing our casualty reporting, postal support, personnel service support, and personnel accountability mission in all of these alphabet locations that were just mentioned in the script. I needed to ensure our junior leaders had everything they needed to be successful. They were all new to their respective area of operations just like me. If I was overwhelmed, how were they supposed to feel? It was a lot to take in during a very condensed amount of time.

Thinking back now I remember also being struck by the sheer number of people who chose to not only attend the TOA ceremony but also stayed to talk and interact long after the official proceedings had ended. The 1st Cavalry Division's Deputy Commander for Support, Brigadier General (BG) Vincent K. Brooks (now a four-star General), my boss and our Brigade Commander (BDE CDR) COL Aundre F. Piggee, several of the other 15th Sustainment Brigade's command teams, and people from throughout the FOB were all gathered around to show their support. I learned that three of COL Piggee's subordinate battalions were headquartered on the Victory Base Complex; the 15th Finance Battalion was on the same Pad as the 15th PSB, and his 68th Combat Sustainment Support Battalion (CSSB), led by LTC Darrell Duckworth, was on Camp Striker.

For me, this was a time to learn. So I sat back and I paid attention. I listened more than I talked and I observed, as new commanders often do.

I guess you could say that in the early days, learning is key. It's truly the central focus of your mission and the Army works to facilitate this process by implementing a very targeted and efficient procedure designed for this exact scenario.

When assuming responsibility for your mission there is a very specific on boarding process you walk through. It's known throughout the military as Left Seat/Right Seat.

The idea is to take time to ensure everyone clearly knows his or her role before the new unit accepts responsibility for the mission. Thus helping to facilitate a smooth transition that allows for continued support and coverage within every single individual position.

When a new unit, or even a new command, arrives to a duty station they will overlap with the departing personnel for a period of at least two weeks. For the first week the incoming Soldier is in the "right" seat, or the passenger's seat. They use this time to observe the departing Soldier and literally shadow his or her every step throughout the day.

After a full week of this, the idea is that the new Soldier has a clear understanding of his or her duties. Not just the big picture job description, but the actual daily steps that make up the role.

Then, about half way through, it's time to switch seats. Now the new Soldier moves from the "right" seat to the "left" seat, or the driver's seat as they switch places.

At this point, the new Soldier is now the one responsible for the daily duties while the departing Soldier becomes the observer. This allows them a full week of one-on-one, on-the-job training from someone who is extremely familiar with the job requirements. The purpose is to allow the new Soldier to actively learn his or her role as someone walks with them through every step, while offering some last minute mentoring.

Once the second week comes to an end, the departing Soldier moves on, and the new unit takes over the command and assumes full responsibility for the mission.

For my Soldiers, this was a very effective approach that helped bring them up to speed as fast as possible. For me, it was a logistical puzzle that was shifting and changing with each passing day. When my unit arrived we took responsibility for the missions of two other units. This meant that, following the TOA, my troops were immediately responsible for providing support for two-thirds of Iraq. A huge undertaking and not one that had been attempted in the past.

I was comfortable with operating within the MND-B's footprint, because this was the 1st Cavalry Division's area of operations. These were the same units the 15th PSB supported during the previous deployment. This was who the 15th Sustainment Brigade was aligned with in the rear at Fort Hood, TX, and this was the same area of operations our predecessors had just supported and were very comfortable with. I attended MG Fil's commander's battle update briefs and was invited to attend any of his commander's mentorship sessions. In short, we knew the MND-B mission was doable. We simply had to learn the systems already in place to support the Baghdad area of operations. The other locations, however, were a much bigger challenge because at those locations the other PSB had already redeployed back to the States, meaning there were no Soldiers to conduct Left Seat/Right Seat with. Instead, we were on our own and we had to figure it out in locations that were unfamiliar to our unit and had unfamiliar leadership. Different branches of our military and different coalition forces also controlled some of those locations, which added to our logistical challenges.

During those early days in theater, there were more hurdles than I can count or even remember. And, while I cannot recount every trial or difficulty we overcame, I can tell you that the entire mission would have been completely impossible without the help of my battle buddy CSM Livengood, and my XO, MAJ Jon Althoff. These men were more than my right hands, they were my team, and I am forever grateful for both of them.

While my description of them can never fully do them justice, I feel it is important to explain them beyond just their names, ranks, and roles within my chain of command. They were so much more than just my CSM and my XO; they were sometimes my advisors and often my confidants.

MAJ Jon Althoff walked and talked with a sense of confidence, loyalty, and humbleness. He was a couple of years older than me and possessed a sharp wit complete with a professional military bearing. He was a devoted husband and father of three young teenagers and had been an NCO prior to commissioning into the officer corps. He had been assigned to the 15th PSB only 3 months prior to leading them into Iraq on this deployment. Jon was rock solid.

CSM Livengood had already been assigned to the 15th PSB for two years, had actively trained, mentored, and counseled a majority of the Soldiers in the unit during their most recent deployment to Iraq in support of OIF II, and he prepared them for this current deployment. He had served with the two previous BCs. He was physically a big man who naturally possessed command presence. He was smart and reliable, yet stern when necessary.

When CSM Livengood and I initially met, I was hesitant to trust him because, even though the previous commander was relieved of his duties, CSM Livengood was still there. Meanwhile, I knew he was sizing me up just as I was gauging his skills and abilities; he had to learn to trust me, just as I had to learn to trust him. It took a few months but then we were like two peas in a pod. We were both very direct in our communication styles and we made a fantastic team because we both expected the best no matter what. Looking back on this deployment, I have a million memories of working with him, but I can remember one particular incident that had us both laughing out loud and ended up setting the tone for our command's interactions with our Soldiers.

I'm not sure how it started, but during our many trips to visit our Soldiers, the junior Soldiers would approach me and open their arms for a hug. And not just a small pat on the back but a giant bear hug that was all encompassing and lasted for several seconds. While hugging is pretty common outside of the military, even in certain professional situations, in the military this type of behavior is generally unheard of. Especially when you are referencing a junior Soldier approaching and hugging a Battalion Commander.

It's important to note that these Soldiers were all out on small FOBs by themselves and they knew the CSM and the BC were coming to see them and that we would do anything for them. However, regardless of the circumstances, CSM Livengood was not a fan of their habit of hugging me. He quickly and loudly voiced his concerns and told me quite plainly that I needed to stop doing whatever I was doing to make them think it was okay to hug me. So I tried to discourage them. It didn't work.

Now he became really determined to stop them. So instead of greeting our Soldiers standing side-by-side, he would stand in front of me and literally try to block the Soldiers from reaching me.

That didn't work either. It didn't even slow them down.

They would just dodge right around him and hug me anyways while I awkwardly stood there without reciprocating the hug because the entire process just didn't seem appropriate by military standards. After this happened several times on visits to several different FOBs, CSM Livengood threw up his hands and gave up. He realized that even though we might not like it, we weren't going to stop the phenomenon, so we might as well get used to it.

The three of us, CSM, XO and myself, made a great team because we were all focused on one mission and one goal. We dedicated ourselves to taking care of our Soldiers and getting the job done, no matter what. We had a great synergy amongst our team and it showed. Not only in how we did our jobs, but in how we led our troops. I am honored and grateful to have served with each of them.

Priority one was to make an assessment of our battalion staff, command teams, higher headquarters, and the overall mission. We had a lot of ground to cover and knew we would need to circulate around our battle space consistently if we were going to be successful. So we created a plan and we got to work.

The first mission I gave my XO was to create a visual tracking system of Task Force Bronco (our Soldiers were known as the Broncos). I wanted to know how often someone from our command visited one of our units in any of the 26 alphabet locations throughout Iraq, and to keep track of our Soldiers' requirements such as equipment, training, and support. This tracking system allowed us to note each individual visit and ensure that we didn't inadvertently skip any location within our designated area of responsibility during our regular rotation. With the sheer number of Soldiers we were responsible for, we had to be very purposeful in our actions to keep our team running effectively and efficiently and to make sure all of our Soldiers were taken care of.

For me, this was my third deployment to a combat zone. I had, however, deployed numerous times beyond the U.S. borders to countries including Kuwait, Egypt, and Saudi Arabia, and within the U.S. borders on field training exercises simulating battlefield conditions. While I was assigned to an infantry brigade in Alaska, I learned one of my most valuable leadership lessons. I found out firsthand that if I wanted to be successful, then I needed to trust my

NCO's and communicate authentically and effectively. During this exercise, the assigned leadership provided little guidance and practically no support or resources throughout a training scenario conducted in negative 50-degree weather. Conditions this harsh can be deadly and when mixed with a toxic command climate it is a recipe for disaster.

At the end of that exercise I knew for sure I did not matter to my chain of command. I also knew that I would do everything in my power to ensure that none of my Soldiers ever felt the way I did during that exercise. I was determined that each member of Task Force Bronco would know they were a part of a team of excellence. And to that end, they would know that their leadership knew who they were and cared about them as an individual. Either me, CSM, the XO, or someone from the battalion staff would visit each of our locations at least once every three weeks.

We were fortunate that our BDE CDR was constantly away from Camp Taji, his home base roughly 30 miles north of our FOB, visiting Soldiers because he and his CSM always stopped by the post offices and pax terminals (specific locations that tracked military personnel as they moved about within the theater and recorded when they left) that we were responsible for in that area. We also had subordinate elements geographically dispersed in the 26 separate alphabet locations within Iraq, plus some in Kuwait and Qatar, and others back at Fort Hood and at Fort Sam Houston in San Antonio. Covering all of that ground on a regular basis quickly became an exercise in logistics. But we made it work.

CSM Livengood and I traveled throughout two-thirds of Iraq to visit our Soldiers and the units they supported continuously. Sometimes we would travel anywhere from Kuwait to Ramadi, Kalsu, Rustamiyah, Bucca and Basra, or roughly 1,000 miles total through a combat zone, all in the matter of a single week. We learned how they performed their mission from the supported elements. We learned how higher headquarters supported them and how they were treated on their FOB.

Some units would allow their Soldiers to help our Soldiers pitch the mail or provide forklifts and maintenance support, while also inviting our smaller elements to their mentoring sessions. This helped to facilitate a strong relationship between our Soldiers and their FOB leadership, and was the kind of environment we tried to foster. For the most part, we were very pleased. However, on some occasions we had to retrain and refocus our junior officers and NCOs on the standard of excellence we expected. On rare occasions we had to engage with the FOB leadership when we observed or were told by our Soldiers that they were not being supported in their respective area of operations. We were less than impressed with the "we" and "they" mentality some immature leaders would adopt. It usually only took our presence and engagement to adjust that point of view. However, on one occasion we threatened to pull the postal and pax terminal support from a specific location due to the treatment my Soldiers received there. This was the wakeup call the leadership needed to embrace my Soldiers and treat them in a fair manner.

The second mission was to have the S1 set up my rating system for tracking evaluations and awards. I intended to counsel each officer, warrant officer, and senior NCO in writing. I learned this technique from my battalion commander, retired COL Tom Owens when I was a 2LT at Fort Bragg, North Carolina. When I served under COL Owens he used our written counseling sessions to give me my left and right limits. This helped me to see how my personal actions, small as they may be, affected the mission as a whole.

Interestingly enough, his biggest impact on my career was when he fired me as his battalion S1 and reassigned me as the XO for the 129th Postal Company. Once I settled into my new role, he would walk across the dirt road to the post office where I now worked and he would sit and talk with me. In those moments I knew I mattered both to him and to the mission. Because of this lesson, I knew that as a Battalion Commander I wanted my Soldiers to know how important they were to me, but I also wanted to make sure we showed support for our front line troops as well. Every Soldier in Iraq, whether under my command or not, was an important part of our overall mission, and I wanted to make sure each and every one knew that.

So, my CSM worked with my XO and our Battalion S3 (Operations Support) to conduct personnel service support "Rodeos" to remotely located units. These Rodeos were designed to provide support for the front line troops without forcing these Soldiers to leave the fight. The Rodeos were extremely valuable to the individual Soldiers, the units and even the war fighting Commanders. As these Soldiers were highly focused on the task at hand, they did not have the time or ability to travel to an Army Post Office or to a Personnel Services Center. So, we brought the Post Office and essential personnel services to them.

During OIF 06-08, my battalion S3, CPT Jeannie Deakyne took the coordination lead. She determined the centrally-located remote rally point for the service, tasked CPT Rodney Davis' Alpha Detachment with the mission, and coordinated with the 15th Finance Battalion's S3. Our Soldiers would travel in teams of at least four and transport a couple of tuff bins. These bins were large, practically indestructible cases, filled with a complete, deployable Common Access Card system (CAC, our military ID card), two laptops, a printer and associated supplies. Once on the ground they had the capability to produce CAC Cards, CAC PIN resets, ID tags, update DD93 (Next of Kin) and Service member's Group Life Insurance (SGLI) forms, and offer promotions assistance. The representatives from our sister finance battalion conducted cash disbursements, pay inquiries, and allotment starts/stops/changes. These missions were combat multipliers. They often eliminated the requirement for units to put someone on the road where the enemy was employing IEDs more frequently during this rotation. They also allowed us to fully complete our mission of providing our troops with the best possible support.

Keep in mind that we were working to accomplish all of this while keeping track of the thousands of moving pieces we encountered on a daily basis, all under a brand new command structure.

During this time the Army was transforming from a division-centric focus to a brigade-centric focus. This change was years in the making. It was necessary because of the ongoing military requirement to fight in the protracted Global War on Terrorism. When I deployed as a 2LT to Saudi Arabia from Fort Bragg in support of Operation Desert Shield/ Desert Storm, it took months to transport our equipment and personnel to the theater of operations, a fact that was simply unsatisfactory to the current leadership. It was unreasonable to believe that the enemy would simply wait patiently for us to arrive and fight a linear war. During previous military conflicts the strategy was developed at the division level, where a two-star General would direct the brigade-level leaders on how to execute their daily missions. This approach was no longer feasible because the enemy was shifting its strategy.

Therefore, something had to change.

This led Army leadership to decide to undergo brigade-level modularity. The concept was to equip a brigade-level organization, commanded by a Colonel with slices of elements from the separate brigades. It was a brand new approach that would enable the modular units to deploy separate from a division headquarters and be successful. The idea being that with this type of structure the Army leadership would become more flexible and proactive when developing strategies to engage the enemy.

The personnel community's response to modularity was to convert to Personnel Services Delivery Redesign (PSDR), which essentially, inactivated legacy personnel services battalions, including our 15th PSB, throughout the Army and re-purposed the structure by embedding it within the newly developed modular brigades. There was a shift occurring within the AG community, which these Personnel Service Battalions were a part of, beginning with the way we described our mission. The terms personnel, personnel administrative, and personnel administration were quickly being replaced with the heading of human resources.

This was a huge paradigm shift. And, in 2006, when I had troops on the ground in Iraq, the Army was at the very beginning of this momentous process. On paper it looked good. But, as with most organizations during times of change, the reality was quite different.

No matter how you looked at it, the pressure was on. And it wasn't letting up anytime soon.

Despite the challenges we faced with a change of this magnitude, the expectation for success was tremendous. The Army's newly coined Human Resources leaders were determined that their new approach would succeed. Our mission was the first time they would test their theories. In that light, success was mandatory. Failure was not an option and would absolutely not be accepted in any capacity.

The 15th PSB was one of two legacy personnel services battalions remaining in a deployed environment at this time. The Army had already placed the new AG positions within the brigade combat team structure. However, they had not necessarily assigned the personnel to the units.

At that point, the fact that we had assumed a larger mission was irrelevant. The point was, we had assumed the mission. And with that, we would find a way to succeed.

Even with troops dispersed across 26 separate locations and spread over a huge geography.

One thing was certain. We *knew* that without outstanding communication we would be doomed before we ever got started. If your unit doesn't have strong leaders who are able to effectively communicate the mission, then you are on a path for disaster. And, in our case, we had to be able to effectively communicate with troops spread out throughout most of Iraq. My job was to make sure my Soldiers understood the mission, had what they needed, and were properly trained. If my CSM and I did not do that correctly, then how could I expect them to succeed?

It's a question I encourage every leader to ask themselves.

Overall I felt, and still feel, that these changes, although difficult at the time, were intended for the better. This new process replaced the legacy PSS structure that required multiple tedious layers of paperwork. Each additional step adds time and bogs down the process. My belief is that we should always focus on getting the job done effectively, but in the most streamlined way possible. That is why innovation is so crucial. Being willing to change processes and procedures can open the door for innovation and improve the status quo.

For the 15th PSB, the change required as part of this process was difficult, but that didn't scare us. We wouldn't shy away from something because it was hard. We would always be willing to push through the process to find out if it was worth it.

By the end of 2006, Operation Iraqi Freedom had entered a new phase. With the execution of Saddam Hussein, the focus of the war shifted, yet none of this impacted my unit's mission. We were there to support the warfighters and that was exactly what we would continue to do until it was our turn to rotate home. I remember watching bits of the news coverage and wondering what our military and civilian leaders would do next.

Then I remember being very sorry I had ever asked that question in the first place.

For us, this was a confusing and sometimes frustrating time. At this point the national narrative at home in the United States was that the war was all but over. To listen to the news you would think that life in Iraq was a cakewalk, a simple trip to the Middle East with no more danger or worry than a run to the grocery store.

Yet our reality was quite different. In our world, people were dying. We were still filling out casualty reports daily and sending home the messages no one wants to receive. The worst part was that some of these deaths came in the form of suicides. Soldiers who literally couldn't take another moment living in war and chose instead to end their time on earth rather than live in hell another day.

Here is what I urge you to keep in mind. The months of October, November, and December 2006 were relatively quiet in terms of U. S. military fatalities. There are "just a few" names on the lists for that time. But every single Soldier matters and they all have families and loved ones waiting at home; people whose worst nightmare is to receive "that" knock at the door or to see their loved one's name scroll across a news screen.

In October 2006, one of our postal Soldiers was hit with shrapnel after an incident of indirect fire. In December 2006, three of our postal Soldiers at FOB Rustimayah earned combat action badges after they were injured during yet another attack on their FOB during which the post office was hit and some mail was damaged.

So even if there is just one fatality or injury, always remember that someone's world just changed forever. And although the number may not be "significant" according to statisticians, it is absolutely life-changing to the people involved.

This is especially important in our world of 24/7 news access. In this day and age, news is reported almost instantaneously, which can result in information being shared faster than it truly should be. In our role we ran into a few situations where our Soldiers' families would hear information on the news *before* they could learn about it straight from their Soldier. This caused extreme stress for both the Soldiers and their families.

When a Soldier deploys, his or her family is planning for them to come home. They are scared and worried from the moment they leave, but they all *plan* for their Soldier to come home, even though deep down they know that might not happen. So when families at home hear about deaths happening throughout the Iraqi theater, they get scared. And because they can't just pick up the phone and call their Soldier, they have to wait. So once casualties are reported, they are either waiting for their Soldier to call and confirm he or she is safe, or they are waiting for "that" call or "that" knock on the door that no one wants. This makes for an extremely stressful way of life for families left behind during deployments.

This type of constant uncertainty led to a sort of frenzied daily ritual. Every day at mail call or during open times for phone calls, all of the deployed Soldiers would try desperately to connect with their family back in the States. Believe it or not this was twofold. First, they wanted to assure their families they were safe, that everything was fine, and that they would return home shortly.

Second, they wanted to reconnect with what mattered. To make sure their loved ones were still there and would still be there the moment they finally stepped off the plane.

Our Soldiers managed this process and worked extremely hard to ensure it was not interrupted. We were also blessed to have outstanding rear detachment leaders who were very involved with our family support group.

To add to the chaos of this process, there was a constant flow of troops in and out of Iraq. During this time all of the various units were coming and going on what felt like a daily basis. Then, just as we all thought things might begin to calm down, the Iraqi Troop Surge was announced and shocked us all.

The main problem was that it was not announced to us.

President Bush announced the Iraqi Troop Surge to the public on live TV. While our Soldiers did not see that broadcast, their families did. In many cases the deployed Soldiers actually learned about the Surge from family members back home before they were officially informed of the new mission by their chain of command.

I can promise you this is not the way you want to find out critical information, especially when that critical information involves an extension to your deployment.

The Soldiers were shocked and their reactions were all over the board. While most took the news in stride, others came to us crying and a few suffered a full breakdown of military bearing.

However, this change didn't just affect us, but all of the deployed forces throughout Iraq. While our extension was hard on us, the most upsetting story I heard in relation to the surge was tied to the 172d Striker Brigade Combat Team, a 4,000-plus-man unit at the very end of their 12-month combat tour. A small number of Soldiers from this unit, about 300 in total, had redeployed back to their home station ahead of the remaining troops. They landed in Alaska and all but a very small handful were immediately sent back to Iraq.

I cannot, and do not want to imagine what they must have felt in that situation.

Although I knew I had to do the right thing and encourage my Soldiers through this shift, I was actually pissed off with the change in our timeline and disappointed in the manner in which the extension was announced. I remember my BDE CDR providing guidance to all of us BCs, encouraging us to circulate throughout our area of operations and connect with our subordinate leaders to ensure they maintained their professionalism and remained focused on the mission at hand. CSM Livengood and I did just that. We kept a stiff upper lip. Understand that while the timelines for most of the active duty units were extended to 15 months, the reserve units remained on their standard 12-month rotations. So we had a mixture of Task Force

Bronco units that were preparing to extend their stay for several months, while the reserve units were preparing to rotate home.

Through all of the traveling, training, counseling and mentoring, I constantly reminded myself that we had an incredible bunch of Soldiers to work with. Included in our ranks were the AG Soldier and Officer of the Year, and the former AG Warrant Officer (WO) of the Year. This distinction means those Soldiers represent every AG Soldier, Warrant Officer, and Officer in the Army, which is a remarkable accomplishment.

In this way we focused on the good. We kept working and doing our jobs, keeping our eyes on the prize even as we slowly marked days off the calendar. For me I felt like we had finally entered the home stretch. Once the initial shock wore off, the Soldiers began to settle into a type of stability that I thought would carry us through the remainder of our tour.

You should know by now this meant I had another surprise in store that I wasn't quite prepared for.

In June 2007, while on my 17-day Rest & Recuperation (R&R) back in Mississippi where I was fortunate enough to be celebrating my son's second birthday with him, I was informed that the Division's officer slate included my XO, meaning he would rotate out of the unit as a new officer rotated in. Just when we thought we would have some semblance of stability, MAJ Jon Althoff, the XO who brought the PSB into theater and initially led the battalion through the relief of the previous BC until I arrived in theater, was departing. My new XO, MAJ Tim Zimmerman was already in Iraq with the 1st Cavalry Division, so the transition was both horribly and mercifully quick, as I had to say goodbye to MAJ Althoff and welcome MAJ Zimmerman. Changing XO's after having such a bonded leadership team for so long in theater was mentally and emotionally taxing.

This road just kept getting longer.

CHAPTER 3

The Calm Before the Storm

"Use each interaction to be the best, most powerful version of yourself."

- Marianne Williamson

It is interesting to note and to see how life can change in the span of a moment. And in fact, it often does.

I mean, we hear about these things, read books about them, and watch their stories unfold on the silver screen. But for many of us we simply haven't experienced what it is like to stand at the crossroads of "before" and "after" and realize that no matter what happens we can't go back. We can't return to yesterday, no matter how much we might want to, and our lives will never be the same.

It's hard. Not just the physical motions of breathing in and out as we keep putting one foot in front of the other, but also the mental struggle of letting go of a past you weren't ready to leave. Being forced into a change that your mind wants to reject, even as your every cell searches for a way to do the impossible and turn back time, is excruciating.

What's worse is that these moments often come when we least expect them. When we think that all is well and that things are moving forward just as they should.

This feeling, this sense of peace if you will, was apparent on a day that will be forever etched into my memory. It is a day, in truth a string of moments, which passes through my mind with startling regularity.

And it started just like any other day. At least, just like any other day when you are living and working in the middle of a war zone.

The day in question? 10 October 2007.

If I had to describe the mood at camp that day it would be upbeat for the most part. The Soldiers were content. They were engaged and excited because one of our reserve units was rotating home the very next morning. They were staged at the airfield in tents, waiting to fly to Kuwait, and then onward to their home state of Florida. Each of them dreamed of stepping off the plane to a hero's welcome as they finally got the chance to hug their loved ones and return to their lives once again. I'll admit that I dreamed of that moment right along with them. And not just my own homecoming, still several months away, but also celebrating their success and their chance to move forward with their heads held high, buoyed by the knowledge of a job well done.

Unfortunately, some of them never got that chance.

When we deployed to Iraq, the 15th PSB had four organically assigned subordinate company and detachment-level elements and four subordinate companies attached from various geographies. There were three U. S. Army Reserve units and one active duty unit, which included the reserve 834th AG Company, headquartered in Miami, Florida, and the 175th AG Company, an active duty unit headquartered in Fort Lewis, Washington. The 834th provided postal service in the Southern quadrant of Baghdad while the 175th focused their efforts in the northern quadrant of Baghdad.

Outside of these subordinate companies and detachments, we also oversaw several separate platoons and smaller elements where junior officers, NCOs, and other Soldiers deployed without a command and control element. This scenario was somewhat like putting a small child on an aircraft by themselves and then hoping the flight staff would ensure they made it safely to their destination. Only in this case the junior leaders were flying all the way to Iraq instead of just across the country.

The consistent newsworthy, positive combat multiplier during the protracted war seemingly spanning throughout my entire 26-year active duty career was that Soldiers loved receiving mail. It is billed as a bright spot in an otherwise bleak existence. And, while that may be true, many people don't stop to think about the huge amount of work and careful planning that goes into making a post office work in the middle of a combat zone.

I knew first hand. Not only from watching my Soldiers, but also from my own personal experience. During my first deployment to a combat zone some 16 years earlier, I served as a postal company XO. This deployment showed me firsthand the countless hours required to be successful in the thankless job of working in a military post office. Yes, everyone wants to receive mail, yet no one really understands what we go through to make that possible. Some are very quick to cast blame when they are on the receiving end of unavoidable circumstances or delays. Then there is the emotional burden of being the link between the fighting forces and their world back home. I know how a piece of paper can bring unspeakable joy or equal amounts of body-racking sobs. And I have witnessed and supported the effects of both for Soldiers I did not even know.

Then there is the administrative stress associated with the position.

Serving as a Certified Postal Worker, brought with it inherent responsibilities and scrutiny from military leaders who understood the value of their Soldiers receiving care packages and letters from their family, friends, and loved ones. Also, we knew that the ability to purchase stamps and money orders in theater offered a huge value to our Soldiers.

The additional scrutiny of these services came from the United States Postal Office. The fact that Soldiers and leaders could be sentenced to federal prison for mishandling postal assets was always in the back of the postal Soldiers' and leaders' minds. And with our suddenly increased workload, the stress level increased sharply. The amount of mail increased significantly during the troop surge simply because of the five additional brigades arriving in the theater and only a few units redeploying back to the States; this meant our workload increased dramatically without any additional personnel to handle the load.

Unfortunately, some of our postal Soldiers earned Combat Action Badges during their 12-month tour, not a distinction I ever want for my Soldiers, but it is an unfortunate side effect of serving in an active combat zone. Every single week we were in theater, one of our postal elements located somewhere within our area of operations received incoming rounds. Our subordinate postal elements were co-located with our brigade combat teams, the Marine elements, and on most of the small outposts throughout our vast area of operation covering.

In particular, the FOB where the 834th AG Company was headquartered faced significant activity during their time in theater. Not only did they have to deal with the reality of indiscriminate and frequent mortar rounds being lobbed into the FOB on practically a daily basis, but they also fought through two specific incidents during June and July of 2007.

The first incident happened when insurgents rammed a truck through the front gate of the FOB, attempting to infiltrate the perimeter. This resulted in the death of one civilian on the FOB. The second incident involved an attempted ambush of the FOB. In both cases the Soldiers on guard were vigilant and took immediate action, thwarting the attacks, and ultimately protecting the entire base. This is just one more reason why I am proud to have served with some of the greatest men and women in the world; they take pride in their mission and go above and beyond to serve with honor each and every day. Yet, this specific time span weighs heavy on my mind even now as I think back to the constant stress that came from being on guard 24/7. And I am certain that I am not the only Soldier who remembers this as a dark, foreboding stretch of time.

But on that day, at least for a moment, we were all focused on the good. With most active duty units having their tours extended while reserve units still maintained their original 12-month schedule, we had a mix of Soldiers either arriving in Iraq, maintaining their current position for an extended period of time, or looking forward to going home the very next morning.

My day started early and it was scheduled to be jam-packed from beginning to end. By 1600 hours, or 4pm, things were still going strong and I was lost in my work when CSM Livengood stuck his head into my office and told me it was time to head out for the awards ceremony honoring the 834th AG Company. I walked out of the battalion headquarters and looked up at the sky, noticing how bright the color was and how clear it seems. It was as if the whole world was actually happy for once.

Moving around the tall concrete t-walls, which were staggered in a box like shape and surrounded the battalion headquarters, I could finally see a formation of Soldiers standing at attention as they awaited my arrival. I remember being shocked by how quiet it was, as though no one even dared to breathe.

To my left I saw my S1, LT Pinzon looking serious and focused. He had two of his Soldiers holding the tray of medals with stacks of green award binders all pre-organized to align with the format of the ceremony. Seeing all of those medals was like a visual representation of everything I knew my Soldiers had accomplished. It made me proud to lead this group and I happily stepped forward and took charge of the formation.

First, I congratulated the Soldiers on a job well done. They had all done their part to serve their country well and I implored them to share their magnificent story when they returned to Florida and to their eventual home of record. I told them how much we appreciated their effort, energy, and dedication during their time in theater, making sure to look each of them in the eye as I spoke. Then, I moved to ask the CSM if he had any remarks or comments, as was our custom during the 20-plus end of tour award ceremonies we conducted together. The theater policy was to ensure that units received their awards prior to redeployment, meaning we would conduct this ceremony prior to any unit rotating home.

I chose to take a picture with each individual Soldier during each award ceremony the CSM and I participated in. My goal was to ensure that each Soldier knew they mattered and that they were worth my time and attention. It was the least I could do. As I stepped in front of SSG Clamens, we both took a deep breath and paused for what seemed like an inordinate amount of time. I noticed a single tear slowly create a path beginning at the corner of her right eye down the side of her face until it finally rested underneath her chin. I can admit that moment made me suck in a deep whoosh of air as I struggled to keep my own composure in front of the troops. My goodness gracious, I can only imagine how much she was anticipating the reunion with her family, especially her small children.

As soon as the ceremony ended, Soldiers scattered throughout the battalion area of operations. Several made their way back into the battalion headquarters while I traveled back to my office to close out the day. All around me I could hear people making plans as to how they would spend their final night on base. Most of the Soldiers were headed to the DFAC; some were specifically going there to have a celebratory dinner with SSG Clamens to wish her farewell prior to her unit's departure from Camp Liberty the next day.

The noise died down a bit as they all moved to the door and out into the evening. The XO called out that he would catch up with them later but wanted to finish up a few things before he left for the day. I finished up a few tasks of my own and then headed out, following my normal routine of walking out of my office to the left, past the CSM's office and the S4 (supplies and equipment). The hallway ended at the mailroom and the S1 shop. Across from the S1 shop was a row of phones, each a hard-wired landline and they were constantly filled with Soldiers straining to hear the sounds of their loved ones back home. It was always a bittersweet sight.

As I doubled back to depart the battalion headquarters through the door closest to my trailer, I saw SSG Graves walk toward the front entryway. He stopped because SFC Williams had brought him a plate of food back with her from the DFAC. She wasn't sure if he had eaten and was looking out for him, making sure he got a chance to eat.

Then, when I walked to the back of the battalion headquarters and passed the S3's office, I heard SSG Hilliard ask the former S1, LT Capetillo, if she wanted to join them for dinner with SSG Clamens. I heard the LT say she might join them later. It was just the normal banter, common for this time of day and nothing worth diverting my path to investigate. I continued to walk forward out of the back door and directly into my trailer without being hailed by anyone. My goal was to at least wash my face before walking over to the gym to attend the Gospel Extravaganza being hosted by the Division Chaplain's office.

In an effort to escape the feeling of being deployed, I looked forward to hearing some reverential gospel music and Bible-based teaching of the Word. Due to the number of people who were expected to attend the Gospel Extravaganza, including singers from other FOBs, the Chaplain's office had decided to move this event from the chapel to the much larger gym to accommodate everyone coming from other parts of the FOB, Camp Victory, and the Green Zone.

As many of us prepared to go to the Extravaganza, SSG Clamens, SSG Hilliard, and several of my other Soldiers were making their nightly trip to the DFAC. These two incredible NCOs who had spent the last 12 months in daily contact with each other both on the job and off, had learned to depend on one another professionally as well as personally. They had grown very close during that time and would often share stories of their children and their lives back home to pass the time. On this particular evening, after the end of tour awards ceremony for the 834th, LT Pinzon was walking with them as they made their way to the special dinner where they planned to celebrate their triumphant accomplishments.

It wasn't a long walk, about 15 minutes and just over a quarter of a mile from our battalion headquarters to their destination. Their path took them right by the gym with so many already inside. I'm sure they glanced at the door and likely smiled, glad that their fellow Soldiers were getting the chance to enjoy themselves before many left the base for the last time.

It was an uncomplicated, truly enjoyable evening.

Today I sit here and I imagine their sense of relief. Their joy at finally heading home and the satisfaction that comes from a job well done. I know the phone calls and preparation I made when I was within 24 hours of departing Baghdad and know that SSG Clamens and other members of her unit likely did the same.

If only that idyllic evening had stayed that way.

CHAPTER 4

When the World Stands Still

"The soldier above all others prays for peace, for it is the soldier who must suffer and bear the deepest wounds and scars of war."

- Douglas MacArthur

When it was time for me to leave for the Extravaganza, I jumped into my personally assigned white SUV in an effort to quickly cross over to the gym. We had appropriately nicknamed the truck my "Bronco" and I loved it! The name was a shout out to the battalion, known during this deployment as Task Force Bronco. We were the 15th Personnel Services Battalion. We were the Broncos.

Army-wide Soldiers use the word "hooah" to greet one another, to show their support, to motivate others, or in endless other scenarios; the word "hooah" is often used as a verb or even as an adjective depending on the particular situation. The Army had recently unveiled the new recruiting slogan, "Army Strong" early in our deployment. In the same vein of "hooah' and "Army Strong", our small, but mighty battalion used the words "Bronco Strong". Whenever we finished a meeting or sounded off in formation, we ended it with a resounding, thunderous, "Bronco Strong!" It was our thing.

Hopping out of the Bronco to walk into the gym, my mind was cluttered with thoughts of my 28-month-old son and my mother. Before going to the Extravaganza I had called my niece who was graciously serving as my son's guardian. I was always so grateful to talk with her, but the moments were bittersweet.

Honestly, when I called I didn't think I would catch her. In Iraq we were 8 hours ahead of Mississippi so it was only 9am or 10am when I had the chance to use the phone. I thought she would be at work and I would leave a message, but instead she picked up. I asked how my baby was doing, like I always did, and this time she had a story to share.

The night before, she was sitting in her living room painting her toenails a bright, vibrant red. My son kept staring at her toes and saying "Mama toes!" My niece was surprised by his words and kept correcting him and saying she was not his Mama. Then, she remembered I usually wear red polish on my toes and keep my fingernails clear to remain within the Army's uniform regulations. It was absolutely precious, but it also broke my heart. The thought of my son, my precious baby, remembering that I wore red nail polish on my toes literally broke my heart in half. I missed him so much I could hardly stand it.

So when I walked into the Extravaganza, my heart was heavy. I dropped down into my chair, an emotional mess on the inside, yet at the same time I was relieved to be in a place surrounded by people who were looking forward to something powerful and positive happening during the Gospel Extravaganza. I was more than a little distracted and agitated when the lay preacher opened the service with prayer. But as I looked around, I realized how the gym had been transformed into a sanctuary and a peaceful place of worship.

For me, it was amazingly heartwarming to see this transformation and to be surrounded by the comfort of believers. During each of my deployments, I made a point to seek the company of those who believe that their God is both a healer and a redeemer while also serving our wonderful country. I have found that biblical principles are a steady source of strength for me during what has proven to be a difficult and highly stressful time.

On this particular night, I was especially glad for the company of my brothers and sisters in Christ. The event was set to feature testimonials of people who had grown in their faith over the last 12 months, while also acknowledging the set of people who had diligently supported the ministry on Camp Liberty.

Looking around the room I saw several familiar faces and even waved a time or two before finally taking a deep breath and sitting back in my chair. I relaxed into the moment and for just a short time let go of the guilt I carried over leaving my son. Instead, I chose to focus on how grateful I was that he was in kind and loving hands. He was safe. He was thriving. He was loved. For today, that had to be enough.

That's when it happened.

I don't really know how much time passed, but suddenly it sounded like the world was exploding. A huge BOOM shook the ground followed quickly by another…and another…and another. It was as if the rounds were moving closer to where we sat with each deafening crash.

Initially I didn't move. Not out of fear, but because after 12 months of living in a war zone, this was a sound we were all too accustomed to. But this time was different. This time the explosions kept coming, one right after another until it seemed like they would never end.

Boom! Boom! Boom! The walls trembled with each strike as the noise continued to increase.

So we ran to take cover. Within just a few short minutes there were over 20 of us huddled together silently in a cement bunker.

Glancing over I could see the next bunker and even through the dim light I could tell it was also filled to capacity. From my position I could clearly see the face of Private First Class (PFC) Yvonne Adams, one of my Soldiers. Her expression said it all.

This attack was definitely different. Then, just as suddenly as it began, it stopped.

The attack ended and the night faded into silence as the rain of mortar rounds came to an abrupt end. Before I could even breathe a sigh of relief, a hand fell onto my shoulder. Turning slightly I could see SSG Brandie McKinley and hear her clearly say, "I've got the BC."

I quickly surmised that she was talking to First Sergeant (1SG) Anthony Williams who, from his post at the battalion headquarters, had begun the personnel accountability process. As the 15th Personnel Services Battalion's Headquarters & Headquarters Detachment (HHD)'s Detachment Sergeant, he was responsible for maintaining accountability, training, care, and comfort to all of the Soldiers assigned and attached to the Battalion's command element. He knew that in the face of an attack it was his role to account for each and every one of our Soldiers. For him, it was a routine he knew all too well.

With SSG McKinley's words and the weight of her hand on my shoulder, my mind began to systematically race as my training kicked in and my feet moved of their own accord. Honestly, sitting here today, I'm not even certain if I waited for the all-clear signal. Based on the timing and general location of the attack, I knew in my gut that we had Soldiers in the path of destruction. So instead of moving back towards the battalion headquarters, I jumped behind the wheel of my SUV and drove into the chaos. At that time I didn't have an exact location in mind. My goal was to move closer based on where I heard rounds strike in order to find the main impact site.

I had Soldiers to find.

As I inched my way forward, I could see what looked like droves of people stumbling and walking dazedly towards my vehicle. They were covered in dust and debris, moving away from the impact site as they made their way in groups of two or three to the Troop Medical Clinic (TMC) which was located near both the gym and the Camp Liberty DFAC. I pulled over and walked to the TMC.

Smoke filled the air, so thick it seemed you would choke on it when you breathed it in.

The sky was no longer blue; it was dusty and smelt of burnt meat. Soldiers were crawling, limping, or managing to walk, moving forward in any way they could in what seemed like slow motion. Several had blood on their uniforms, although I could not tell the extent of their injuries. They instinctively formed buddy teams and most of them had someone supporting them.

The first person I recognized was CPT Rodney Davis, my Alpha Detachment's Commander. He was searching through the crowd, frantically moving from person to person as he looked for faces and names and immediately reporting them back to his Detachment Sergeant, SFC Scottie Jarrett. They had also recently released their Soldiers for the day, many of whom were in the vicinity of either the living quarters on Pad 11, the gym, or the DFAC. Some remained in the office to finish out their work for the day, but the majority were spending their personal time however they pleased, meaning many were in the direct path of the attack.

CPT Davis and SFC Jarrett both looked extremely physically fit, were technically competent, and they both exuded confidence; their Soldiers were proud to be in their unit. This was CPT Davis' second combat deployment. While he and I were close in age and had served in the Army for about the same length of time, our career paths were very different. He had a unique military perspective because he was currently serving as a prior service officer, meaning he had served in the Army as an NCO prior to being commissioned as an officer.

I was sure he had witnessed many more chaotic scenarios than this and I trusted him implicitly. During the time we worked together it became clear that CPT Davis only needed to know my intentions to do his job effectively. He and his Detachment Sergeant always conducted thorough analyses, which led to sound decision-making. He never brought me a problem without sharing a few possible solution sets. That night he was certainly serving as a source of strength by maintaining a calm demeanor throughout the evening. If he was as rattled by the day's events as the rest of us, he didn't let it show. Not even for a second.

Accountability was in full effect. I stood in the midst of what seemed like ants after you disturb their anthill. There were people everywhere. Some wounded and looking for help, but more already working to right the wrong. All around me Soldiers of all ranks were moving quickly, helping, cleaning, and lending a hand wherever it was needed. It made me proud once again to serve with such an amazing group.

All of these images, these horrors were flooding me at the same time. It was hard to focus and even harder to look around and see so many Soldiers needing help, so many with wounds that must have been excruciating, and yet they still kept moving forward. I found myself mentally searching for every injured Soldiers' battle buddy. They each seemed to have someone close by, while I continued to remain focused on accounting for my troops. In the end I had to force myself to keep going and to trust that the Soldiers were helping one another to the TMC and that when they arrived the medical personnel on Camp Liberty would be up to the challenge. The best way I could help in that moment would be to do my job and find my Soldiers.

So I kept going.

Before long, night began to fall and I still had not accounted for all of my Soldiers. There were still far too many names missing.

CPT Davis shouted, "Ma'am there is the S1." I looked up and saw LT Cristian Pinzon. I was elated to see him, however, the look on his face was one of discomfort. He attempted to match my enthusiasm, but he could not. I met him, embraced him, and patted him on his back, not knowing what he had just experienced. Not knowing that he was in physical, mental, and emotional pain. I was just so thankful to see him and to know he was safe. His was now one more name I could cross off my still too long list.

After releasing LT Pinzon and speaking again with CPT Davis, I ensured he was in contact with the battalion operations cell and Camp Liberty's TMC. Everyone was following protocol to a "T," but time kept passing. And with each minute, names that were left unaccounted for, the sense of dread in the pit of my stomach grew and clawed its way up to my throat until I had to fight to keep it contained.

Then I saw SFC Elemenia Malone, she was assigned to the 1st Cavalry Division's Special Troop Battalion as the maintenance NCO, so she knew several of my NCOs. Many of them were dual-military service members like her, meaning their spouse was also serving our country, either in Iraq or back in the States. Their kids were resilient and were used to moving every two to four years because of their parents' career choices.

Although she was assigned to another unit, SFC Malone spent a great deal of time with my Soldiers and even attended the majority of our team-building events. At this point we viewed her as an adopted member of Task Force Bronco. Seeing her now, I could immediately tell something was very wrong. She looked strange, visibly shaken, and she seemed to be trying to tell me something but the words wouldn't form coherently. I asked her repeatedly if she was ok, but she wasn't able to answer even though she tried.

Knowing she was in no shape to go anywhere on her own, I told her to sit in my car. She walked slowly to my white SUV and got in, almost as if on autopilot.

It wasn't until a few years later that she shared with me what she wasn't able to say at that moment. At the time the reality of what she had just experienced was simply too much to wrap her mind around, much less communicate.

When the attack first started, SFC Malone was standing beside one of concrete tall walls (t-walls) surrounding the Camp Liberty DFAC. She stood, frozen in place and not sure where to go, until the rounds finally stopped falling. Once the quiet descended, she felt a sense of heavy dread. Her adrenaline kicked in and she ran to the front of the DFAC. Skidding to a stop she saw both uniformed and civilian men and women laid out in a line near the front entryway into Camp Liberty's DFAC. They were obviously injured, or worse, and other Soldiers and medical personnel were already moving to render aid.

SFC Malone, as she began to work through her own accountability process, suddenly recognized two familiar faces. She could remember hearing herself call out to SSG Hilliard and then running over when her friend motioned. She could quickly tell that SSG Hilliard was in a great deal of pain. Falling to her knees she gripped her friend's hand, bending close to talk with her and encourage her through this difficult moment.

As soon as the medical team arrived on site, SFC Malone frantically called them over to help SSG Hilliard. It wasn't until they forced her to move aside that she realized SSG Clamens was lying right next to them, her legs actually entangled so tightly with SSG Hilliard's that the team was forced to separate the two before they could move both of them to the medical truck.

In the truck, they placed SSG Hilliard on the right and SSG Clamens on the left. SFC Malone noticed that SSG Hilliard was not fully awake, but she could still see that her eyes were open and darting around to take in her surroundings. She also noticed that SSG Clamens seemed to look straight through her and was not moving at all. It is an image she will likely carry with her for the rest of her life.

SFC Hilliard would not let go of SFC Malone's hand. Once they arrived at the TMC, the medical team eventually pried their hands apart.

After settling SFC Malone into the passenger's seat in my Bronco, I still was not sure where to continue looking for my Soldiers but I decided to return to the nearest clinic to see if anyone had any new information now that a little more time had passed.

When I arrived back at the TMC, CPT Davis met me almost before my vehicle came to a complete stop. He led me into the building and I could hear him tell someone, "This is my BC." I couldn't tell to whom he was speaking. I only know my heart was racing and I was hoping that this nightmare would soon end. We walked straight through the small TMC and finally ended up outside again. My eyes were fixed on a black bag lying in the rear of the clinic. At that point I think I already knew, I just didn't want to believe it. But either way I had to verify the truth for myself. It wasn't enough to take someone else's word for it.

I asked them to unzip the bag. It seemed like the noise the zipper made pierced my ears. It was noisier than I expected.

Her braids came into view first. Then her face. It was true.

I closed my eyes, bowed my head, knelt down beside her, and prayed right in that moment. I prayed for her family, her kids, and the other members of her unit, many of whom were already in the tents at Baghdad International Airport (BIAP) preparing to fly out the next day to Kuwait and onward to their home station in Florida. Her life mattered. The reality of being here all came into focus. In my mind, at that very moment, I could picture her kids being inconsolable when their father sat them down to tell them that mommy had gone to heaven.

I could have stayed there with her longer, I wanted to, but I knew I still had a Soldier unaccounted for, so I had to keep looking.

At this point, approximately 45 minutes had passed since the barrage of booming sounds had shaken the ground and I still had not found SSG Hilliard. I also needed to make contact with CSM Livengood to update him on our Soldiers. By the time I got back to my vehicle, SFC Malone was no longer inside. People were everywhere as the TMC staff had quickly established triage stations out front to handle the number of incoming patients. Taking care to weave around everyone, I drove back to my battalion headquarters physically alone, and even more distressing, alone with my thoughts. I had to focus; I assured myself there would be time to mourn later. Right now, I was looking for my CSM and SSG Hilliard.

CPT Seth Graves met me at the front door of the battalion headquarters. Although he did not have any answers in terms of SSG Hilliard's location, he did offer to drive me across the base to Camp Victory's TMC.

It made sense for CPT Graves to help with the search, as SSG Hilliard was assigned to the battalion S1 staff section that he was in charge of. She served as the Non-Commissioned Officer in Charge (NCOIC) of the personnel section. CPT Graves got in the driver's seat of a nearby vehicle as I opened the passenger-side door. Now, looking back, I imagine he knew I was a mess and he was attempting to do the right thing for the unit, despite his own fears, as we looked for SSG Hilliard. It is a gesture I appreciate to this day.

As we arrived at the Camp Victory TMC, I noticed that it was pretty quiet on the outside compared to what I had witnessed at the Camp Liberty TMC. The inside, however, was bursting with people and activity.

A few feet inside the front door, I stopped someone in medical fatigues and asked about SSG Hilliard. In this type of attack the medical teams simply start working to offer help as quickly as possible. They start with the most severe injuries and then continue working until every Soldier has received care. Due to the frantic pace, they do not always know which Soldiers are where at a given time. So, while he went to check on the status of SSG Hilliard for me, I sat down to wait beside a Private I had never seen before.

Looking over, I noticed that his hand was extremely swollen, it looked almost inflated as if it were a balloon. It was striking because he sat so quietly and calmly, yet was obviously in extreme pain. Shifting to the side, I asked him if anyone was helping him. His response made me catch my breath. Maintaining perfect military decorum he answered that he was not as injured as others and could wait. He then mumbled that he also did not know how to get their attention.

In disbelief, I stood, looking straight at the young Soldier and told him to come with me. I walked him to one of the medical personnel and demanded that someone take care of him immediately. The Private looked back at me as he walked, what seemed to me to be much too slowly, and said, "Thanks Ma'am." Again I was struck by his ability to remain calm and disciplined while in so much pain.

Another true hero.

Moving back to my chair, I sat down again to continue waiting for news of SSG Hilliard. I watched as each of the beds filled with Soldiers injured in the mortar blasts. I noticed BG Vincent K. Brooks actively engaged in providing medical support. If I did not know better, I would have thought he had previous medical training as he assisted alongside the medical personnel. I noticed blood on his Army Combat Uniform (ACU) jacket and felt reassured that the Soldier he was helping would surely be fine. The Soldier was conscious, yet seemed to be seriously injured.

As I sat and not so patiently waited for news of my Soldier's condition, I surveyed the room. People were rushing here and there to help the wounded, while blocking everything else out. The smell was strong and strange, not something I can easily describe, but something I have not forgotten in these eight years. It was a mixture of dirt, sweat, and blood. Even today if I smell something even close, even for a moment, my stomach begins to roll.

A member of the medical team approached me and I stood to focus on their words. My stomach clenched as they informed me SSG Hilliard was not at the clinic.

I still did not have 100% accountability.

Leaving the clinic, I finally saw CSM Livengood for the first time since the attack. As I walked outside he was climbing out of his SUV and moving quickly towards me. As our eyes met, he sadly shook his head, telling me SSG Clamens did not make it. Even though I already knew this, his announcement made the stark reality that much more real. I fell to my knees right there in the parking lot of the TMC as we both sobbed in the night.

The wave of emotion rolled through me, not just for SSG Clamens, but for the full pain of the night's atrocities. As a Soldier, you are trained to put emotion aside and act, to move, almost as if on autopilot, through your tasks until the mission is accomplished. But, emotion can only be held at bay for so long.

Once I was able to regain my composure the CSM informed me that SSG Hilliard had been medevac'd to the 28th Combat Support Hospital (CSH).

100% accountability achieved.....finally.

With that, we got into his vehicle to drive back to the battalion headquarters together. There was still work to be done.

Later, I found out that SSG Clamens was not our Brigade's only casualty of the night. Our sister battalion, the 15th Finance Battalion's Soldier, Specialist (SPC) Pearson, also lost his life during the attack. I later learned that SPC Pearson had only been in theater for a few weeks.

Right now I know that some people will read this and say, "It was just two." For me, it's hard to even have the words to respond to this. Two may seem like a small number, but one was *mine.* It was my job to bring them all home, and I wanted and expected to do just that.

Any casualty is unacceptable.

CHAPTER 5
A New Reality

"In war, there are no unwounded soldiers."

– Jose Narosky

As we pulled away from the TMC, I remember staring blankly out the window in disbelief. Silence hung heavily in the air as we both struggled to come to terms with this new reality. CSM Livengood seemed lost in his own thoughts, and in that moment we were each just trying to accept the truth of the last hour.

For each of us, this was our third deployment to a combat zone. About 16 years prior, we both participated in Operation Desert Shield/Desert Storm at the beginning of our military careers, albeit in very different units and very different capacities. My second combat deployment was in 2003 as an individual augmentee with the United States Central Command (USCENTCOM or CENTCOM) in support of Operation Enduring Freedom. CENTCOM's main headquarters is located in MacDill Air Force Base in Tampa, Florida. I deployed to a forward headquarters established at Camp As Sayliyah in Doha, Qatar in 2003 directly from Fort Carson, Colorado.

CSM Livengood's second combat tour was as the CSM of the 15th PSB in OIF II, following his successful completion of the United States Army Sergeant Major Academy's rigorous 10-month course in 2004. Yet, even with all of our combined experience, this was the first time a Soldier under our direct command and control was seriously injured, much less killed. In CSM Livengood's previous deployment during OIF II, there were no significant physical injuries to the 15th PSB. At that time, the PSS mission was pretty much the same, however, not as expansive, meaning his role during this tour was very similar to his earlier experience. Therefore, it makes sense that he would expect to not only be able to lead his Soldiers, but also to successfully bring them all home.

CSM Livengood was a big man in both stature and personality. He had a memory that was incredibly clear and he could rattle off substantive statistics quicker than anyone I have ever served with. He knew how many pounds of mail our Soldiers were processing at any time during the deployment. He knew how many units we were supporting and their exact locations. He could tell me the command teams' names prior to our visit, no matter how remote the location. He always took great care to ensure I had valid talking points as we travelled extensively together throughout two-thirds of Iraq, conducting battlefield circulation, and engaging with the 15 or so subordinate units that our battalion had the honor of serving as their command and control element. Training, mentoring, and counseling were his forte. He was both feared and respected by our Soldiers and the junior leaders assigned and attached to our unit. I relied on him greatly.

As we arrived back at the Battalion Headquarters on PAD 11, we found the battalion S3, CPT Jeannie Deakyne, waiting for us. She met me at the door and briefed me on the current situation, outlining SSG Hilliard's condition and informing me that she had arranged a Blackhawk helicopter to fly myself, the CSM, and the HHD Command team, CPT Seth C. Graves, and 1SG Williams, to the International Zone (Green Zone) within Baghdad so we could have the opportunity to see our Soldier.

Throughout the course of my military career I sat on countless flights, yet none stand out quite as clearly as that one. Even with the hundreds of flights I utilized during that deployment, I can't remember ever being more on edge as the helicopter took off. It is interesting because we often used travel time as an opportunity to take a quick nap to catch up on must needed rest. This was not one of those flights. Instead I stared out into the darkness, scanning for the telltale flashes sparking in the night that would indicate incoming fire.

As we flew through the night, I couldn't help but wrestle with the reality of the last few hours. My mind was spinning and the "what if" game was ceaselessly haunting me. One at a time, pictures flashed through my mind of each of my Soldiers who earned a Purple Heart during our deployment. Then I focused on those who earned combat action badges. Every single one of those Soldiers lived. They survived and they would have the chance to go home to their families.

Next, my mind drifted back to one of the many mortar attacks that wounded my Soldiers, yet they were still able to return to duty. Such as the three postal Soldiers from Montana who supported FOBs Rustimayah, Loyalty, and Hammer for 12 months and processed nearly 2.4 million pounds of mail while supporting over 15,000 Soldiers.

That post office took a direct hit.

The more I thought about each of these situations, the more anxious I became. I thought about our casualty liaison team (CLT) and all they had witnessed. All of our PSS missions are important, however, the casualty mission is one of zero defect. Task Force Bronco had five CLTs located throughout our area of operations: one each embedded with the 1st Cavalry Division and the 3rd Infantry Division, and three based in hospitals including one team at the 28th Combat Support Hospital. The CLTs were an essential part of the Department of the Army's overall casualty operations mission; to record, verify, and process casualty information from the unit level to the Casualty and Mortuary Affairs Operations Center, notify appropriate individuals, and provide casualty assistance to the Next of Kin (NOK). This small five-person team was an intricate part of a huge web of individuals who were producing, disseminating, coordinating, validating, and synchronizing casualty reports.

The teams consisted of five Soldiers who did much more than their doctrinal role. They reviewed Defense Casualty Information Processing (DCIPs) reports all day during their 12-hour shifts. They were responsible for reaching out to the respective divisions' subordinate units to consistently gather additional information on a particular incident. When the Soldier was Killed in Action (KIA), Dead on Arrival (DOA), or Returned to Duty (RTD), the CLT followed the administrative trail to the end. Those embedded with the Division Headquarters had to read about their fellow Soldiers being hurt and killed day-in and day-out, but they did not have to see the casualties face-to-face; those located in the hospitals, however, saw each and every casualty that came through their facility.

Once a casualty arrived at the combat hospital, a CLT member interviewed the patient to get his or her vital information. This information was then entered into the Defense Casualty Information Processing System (DCIPS) and transmitted to the various higher headquarters confidentially and efficiently. While the patient was in the hospital, a CLT member entered a progress report every 2 hours. Once a determination was made as to whether the casualty would be medically evacuated or returned to duty, the CLT member submitted an additional report and contacted the Soldier's unit to coordinate pickup of the Soldier.

We intentionally rotated the teams every 90 – 120 days during OIF 06-08 in an attempt to prevent combat stress because this is understandably an incredibly difficult assignment. Having to face the injury and death of your fellow Soldiers day after day is a heavy burden to carry. My Soldiers were dedicated and they completed their mission with pride, but I felt, and still feel, that none of them deserved to face that particular role for an extended period of time.

All of this played on an endless reel inside my head until the moment the Blackhawk finally touched down and we were able to make our way to the combat hospital.

When we arrived at the 28th Combat Support Hospital, our CLT informed us that medical personnel were attending to SSG Hilliard just around the corner from the entryway. The NCO spoke to us directly, then pointed to a bench nearby. As he kept speaking with CSM Livengood, I walked towards the bench. Looking back over my shoulder, I saw CSM Livengood motion in my direction and then point towards what looked like a shower curtain hanging across an opening in the wall. Needing a few extra moments before seeing my Soldier's condition first hand, I continued over to the bench and sat down, still lost in my own thoughts and processing through the aftermath of the attack. I kept going over the accountability process, checking and double-checking in an effort to ensure we had covered every detail and accounted for every single Soldier.

With so much controlled adrenaline and anxiety running through my veins it was difficult to sit still, especially after what seemed like the longest helicopter ride of my life and with my injured Soldier finally close by. It took me less than a minute to grow restless and pop back up to pace within the small space. I was composed as I walked over to the edge of the curtain the CSM had pointed to and carefully pulled it back to see a flood of controlled activity taking place just on the other side.

As I began to watch in utter amazement, I found that I could not look away. From my vantage point behind the curtain, I watched as the medical team actively worked to help my Soldier. I could clearly see her; she was directly in front of me. Combat hospitals are set up for efficiency, so there aren't hallways filled with individual rooms each housing only a single patient. When you are in the field, you typically have a handful of larger rooms, each lined with open beds. This allows the team to move from patient to patient very quickly as they work to help each Soldier and treat all types of injuries. It was something I was familiar with after all my years in the Army, but something I rarely witnessed firsthand. Unfortunately, today I had a front row seat. My Soldier was lying there on the table while I watched the surgeons' motion back and forth between each other, clearly discussing the procedure and what action they would take.

It was clear they were attempting to save as much of her leg as possible. First, the surgeon closest to me, touched her leg above the knee, then he moved his hand below her knee. They talked about each location and carefully evaluated the situation. From my angle I could see that her ankle was completely gone. There was nothing left there but skin. Her foot was hanging from her body by a thin piece of skin. Even without medical training I could tell there was no way to reattach it.

What struck me in that moment was that the medical team cared a great deal about SSG Hilliard. She was not just another injured body to them; she was a Soldier who deserved their best. They knew she would survive and their actions, their competency, gave me hope that she would be fine. The fact that my mother was an amputee strengthened my belief that SSG Hilliard would go on to live a normal, healthy life. Her road ahead would be challenging to be sure, but, she would be alive.

Once the surgery was finished, a member of the medical team rolled her out of my line of sight. A doctor approached CSM Livengood only to have him point at me. As he walked towards me, I could feel the knot that had formed in my stomach. I hung on his every word. I was grateful for the positive prognosis he shared, as well as his attention to detail. He informed me that SSG Hilliard was stable and would remain at this facility for the next 24 to 48 hours. Then, she would be transported to Germany, then on to Washington, DC before traveling back to Texas to fully recover. I thanked him for all he had done. In that moment I was immensely grateful not only for the skill of the medical team, but also for their integrity, and for the respectful way they treated my Soldier.

After taking care of a few more details, it was time to board the Blackhawk for a second time that night and fly back to Camp Liberty. It was after midnight by the time we landed, but the base was still highly active.

Upon our arrival, we learned that SSG Hilliard's Soldiers had initiated a DCIP report to the 15th Sustainment Brigade's S1 on their two bosses, SSG Hilliard and LT Pinzon. They also submitted a DCIPs report on SSG Clamens, who they interacted with for the previous 12 months, along with approximately 30 other members of Task Force Bronco who were either in or around the DFAC when the mortars hit. SSG Hilliard processed our previous casualty reports; obviously she trained her subordinates to continue the mission in her absence.

She trained them well.

I can tell you now it was a seemingly endless and grueling night, one with memories that continue to intrude on my thoughts with routine frequency.

CHAPTER 6

So Many Sleepless Nights

"When one person is missing the whole world seems empty."

– Pat Schweibert

After that horrific night I pretty much stopped sleeping. It wasn't a conscious decision. I mean, it's not like I would lie in bed at night and force myself to keep my eyes open. Not even close.

Instead, it was as if I came to fear the memories and the survivor's guilt that now lived inside my own head. I began to dread the night. I hated watching time tick past, knowing eventually the moment would come when there was absolutely no more work to be done for the day and I would be left with nothing to focus on but my thoughts.

And at that point, my thoughts were dark. Very, very dark.

Truth be told I am still struggling to grasp a peaceful nights rest even now, nine full years later. Because every single time I close my eyes I am immediately transported back, back to the absolute worst day in my military career, likely in my life, and there is absolutely nothing I can do about it. The images I see in my conscious mind and in my nightmares haunt me regularly, and the survivor's guilt and the 'what ifs' nearly drive me mad.

I think of my son and wonder why I was blessed enough to make it home to him while SSG Clamens' children now face a lifetime without their loving mother. I can honestly say that attending the Gospel Extravaganza saved my life; stepping foot in the gym that night, instead of going about my normal routine and heading to the DFAC after my daily work had been completed, is the reason I was not with so many of my Soldiers outside the DFAC, directly in the path of the mortar attack. When I came to that realization the day after the attack, it was extremely difficult for me to wrap my head around, and, to be honest, it is not any easier for me to comprehend even today, so many years later.

The day after the attack was when it all became unbearably real. It wasn't about accountability any more, or even about bringing the base back to life after such a devastating blow. Suddenly, we all had to challenge our own mindset. It was as if our paradigm suddenly shifted and we could no longer believe the lies we had told ourselves to stay sane during the year we had been in country.

As AG Soldiers, we could no longer buy the myth that we were safe as long as we stayed behind the lines within the confines of a fortified, fixed location. Because, believe it or not, that is exactly what we told ourselves; that our roles were safe. That yes, we were deployed as part of the war effort, but we were not actually part of the war. Instead we were there to support our warfighters, to do our jobs, and then go home. We were support personnel. At least, yesterday we were support personnel. But today? Today we were Soldiers who had seen combat firsthand, and none of us knew quite what to think about that.

Sure, our mindset may seem somewhat strange. How could we go to war and yet believe we were safe? Trust me, I understand why that would seem confusing. But, this exact battalion was on its second deployment in the exact same roles. Not only were these experienced Soldiers, they had already accomplished this mission once and with no casualties. So, the idea that they could do so again was not so far-fetched. Yet what none of us anticipated was the fact that we were going back to a different war. The environment in Iraq changed significantly between those two times my Soldiers were deployed, and while our mission had not changed, the war had. This time we faced a much more hostile reality and it was something none of us were quite prepared for.

So we didn't think about it. Instead, we worked.

We let go of the national headlines suggesting that only frontline Soldiers were dying in the Global War on Terrorism. We closed our minds to the general commentary and focused on our mission and on making sure we all got home safely to our families. For those last months of our mission, it was one foot in front of the other, one day at a time as we got the job done and kept ourselves as busy as possible.

In the days immediately following the attack, I remember taking a call from my boss, my brigade commander, Colonel Aundre F. Piggee. His brigade headquarters was located on FOB Taji, which was a short helicopter flight away. Although I attended his daily brigade update briefs via CITRIX, a secure teleconference platform, we only met face-to-face about every 30 – 45 days or so.

The scope of his responsibility was expansive. He was in command of eight battalions, all of which conducted road convoys except for his Personnel and Finance Battalions. After the attack, I was so consumed with the mix of gaining accountability, losing SSG Clamens and seeing SSG Hilliard that I had not contacted him. This was a crucial mistake on my part because I always expected my subordinates to keep me informed of the good, the bad, and the ugly. And this was definitely the ugly part of war. His voice was stern, yet compassionate, and incredibly supportive. He provided guidance on how to move forward in honoring SSG Clamens and remaining connected to SSG Hilliard throughout her recovery. Then, he stepped back and reminded me that my mission wasn't over, that all of my Soldiers needed me at this moment.

And he was absolutely right.

As a leader I had to stay focused on the fact that my Soldiers were looking to me for guidance during a very confusing and difficult time. If I fell apart, then what were they supposed to do? Or what if they followed my lead and panicked? It wasn't an option. This conversation with Colonel Piggee forced me to take a step back and look at the big picture. The reality was that we were not going home any time soon. So we had to find a way to not only survive, but to successfully complete our mission each and every day. And that success started with me.

I appreciated Colonel Piggee so much. Not just as a leader, but as someone who took the time to help me focus and to remember that even though I was hurting, others were too.

Shortly after that meeting, CSM Livengood informed me that he was returning to the 28th Combat Support Hospital to be with SSG Hilliard. His plan was to take her Soldiers with him, including her closest friend in theater, SSG McKinley, so they would have the opportunity to see for themselves that she was okay, while also saying goodbye before she was transported back to the United States.

SSG McKinley was a newly promoted Staff Sergeant who always displayed a positive, can-do attitude. She would go above and beyond to support the Soldiers who were in the fight. Her goal was to do everything she could to make their time at war easier and she took her job very seriously. In fact, she often returned to her work area to help a Soldier who called our after-hours number. And, not only would she return to the office, but she would also convince someone else to go with her. At that point we had instituted sign out procedures along with the buddy system, so if she wanted to sign-out, someone else had to go too. It could be a cumbersome process, but to her it was worth it as long as she could do her part to make life in the desert even just a little bit easier.

A few months prior to this particular attack on Camp Liberty, SSG McKinley was moved to the 1st Cavalry Division Headquarters to work on the transactional portion of casualty reporting. She worked 12 hours on and 12 hours off every day, but often worked over to be sure everything was taken care of properly. She read the gory details of Soldiers killed in action, burned in their vehicles, and even those who died as a result of apparent suicide day in and day out. Now, she was faced with the fact that she would read about her best friend in the theater, her roommate during this deployment; which was her second tour of a total of four she completed prior to retiring from our great Army. She was angry, sad, and tired of being in this God-forsaken place. No one could blame her for that.

Before the CSM left the base to visit SSG Hilliard, one of the NCOs who worked diligently within the Battalion S4 shop, SSG Graves, approached me in the hallway. For some reason he was not one of the Soldiers chosen to visit SSG Hilliard, but he wanted to be included. He began to plead his case right there in the hall, asking for my permission to join the group going off base and he made several valid points.

He was her squad leader and had served with her since she first arrived at the 15th PSB four or five years earlier. I, quite frankly, could not understand why he was talking to me as CSM Livengood better served the issue. I also didn't understand why CSM Livengood did not speak up, but to me it seemed like SSG Graves needed closure. He needed to see her. From my perspective, this was a NCO I trusted and who consistently demonstrated professionalism and honor. So, I listened to his points, looked at the CSM and simply said, "He's going with you."

Then, I walked away.

After turning around and walking back into my office, I remembered the guidance from my brigade commander to focus my attention on all of my Soldiers. This meant there was still a very important task left to complete. With a heavy heart, I called CPT Rodney Davis to my office. After speaking with the CPT, I directed him to fly to Kuwait and properly inform SSG Clamens' unit of her death. They deserved to know, and they deserved to find out directly from us rather than hearing the news secondhand.

After CPT Davis left to complete his mission, I requested that someone from the battalion operations cell bring SSG Clamens' former commander to my office so I could personally inform him of her death. Again, he deserved to hear it from me.

The hits just kept on coming. It wasn't enough that I lost my Soldier, feeling that pain rip through me as I dropped to my knees after seeing her lying in a body bag when she should be on her way home to her children. No, I was forced to relive the pain and absorb the fresh wave of grief that came each and every time we informed someone new of her passing; the same was true for all who knew her and cared for her.

For each person we had to tell, it was his or her first time experiencing the pain, which meant that for us, it was just like the first time all over again.

Through all of this, our mission did not stop. There was still mail to deliver. There were still files, documents, and requests to process. None of this stopped so we could grieve. Nor did the expectation that we would take care of all of our other Soldiers who were geographically dispersed throughout the country.

So we worked and we grieved.

General David H. Petraeus, the Commanding General of Multi-National Force – Iraq (MNF-I) presented SSG Hilliard with her Purple Heart while she was in her hospital bed at the 28th Combat Support Hospital. I was pleased that the General took the time to personally deliver her award and to speak with her. It also seemed fitting that she received her award before traveling home to the States. But even with all of that, the Purple Heart is never an award that I want for my Soldiers. I wish it was something that none of them ever earned.

Prior to this horrific attack, Task Force Bronco had three Soldiers earn the Military Order of the Purple Heart, which differs from other military decorations because an individual is not recommended for the award, instead they are entitled to the decoration as a result of being wounded or killed in combat action.

SSG Clamens and CPT Pinzon also earned a Purple Heart. One moment that stands out almost above all the rest came when CPT Davis called the next day to inform me that he had arrived in Kuwait and had carried out the dubious task of ripping the hearts out of the Soldiers in SSG Clamens' unit. He experienced their true raw emotions, fear, anger, and sadness first hand, and carried their grief right along with them. He told me that this was one of the hardest tasks he ever faced during his military career and I fully believed him. I could tell this moment weighed on him, but there wasn't much I could offer in terms of support. I was still too raw myself to help anyone else heal at that point. So instead, I encouraged him to be safe as he made his way back to Iraq.

After all, we still had work to do.

I was grateful for the newly arrived combat stress team that was occupying a space behind the Camp Liberty TMC. My XO arranged for them to come over to Pad 11 and speak to all of the Soldiers who were at the DFAC specifically and anyone else who wanted to attend the briefing was granted access. I remember talking to one of the stress counselors. I initially did not know what benefit they brought to the table and I was skeptical that they would launch into a koom ba ya type of moment and instantly become discredited by my Soldiers. It was, however, refreshing to learn that they were specifically trained to deal with Soldiers in a combat environment who had experienced traumatic events.

What made me skeptical at first soon became a type of lifeline for some of my Soldiers who needed a professional to help them process the realities of that awful day.

Over the next few days, I attended the memorial for the Soldier assigned to our sister battalion, the 15th Finance Battalion, SPC Pearson, who was killed in the same incident on 10 October 2007. Thankfully, my BN XO led the planning efforts for SSG Clamens' memorial.

CHAPTER 7

Remembering What Matters

"The battle, sir, is not to the strong alone; it is to the vigilant, the active, the brave."

-Patrick Henry

When most people think about being deployed, specifically those who have never actually been deployed to a forward, hostile, combat environment, they tend to focus on and talk about the actual business of war, they talk about patrols, firefights, midnight raids and IED's. And, they wouldn't be wrong to discuss them. Those types of high stress, and even hostile, interactions are a definite part of the war effort and something every single Soldier is highly aware of.

However, what they don't tend to think about is the fact that we are actually still *living,* even though we may be deployed at the same time. As such, we are making plans, interacting with friends, and even taking part in events as part of our regular schedule.

When I deployed to Saudi Arabia early in my military career as a 2LT, my mother sent me silk pajamas, vanilla pudding cups, and Vienna sausages in one of the many care packages she and my family mailed during that six-month combat tour. In the short note she included inside of the box she wrote something similar to 'I know you are living in hot, nasty tents and sweating all day, but remember you are still a strong and beautiful woman of faith.' Because of that, I made it a point to always have a way to connect with others and to my faith during my multiple deployments and overseas trips.

As a unit, we planned, participated in, and hosted more events than I can even count. Most of the events took place on Victory Base Complex simply because that is where the bulk of Task Force Bronco lived and worked. CSM Livengood and I attended 99% of the events that our outlying subordinate elements hosted, thanks to the Battalion S3's meticulous coordinating efforts. From the TOA ceremony in November 2006, when we assumed responsibility for the PSS mission in two-thirds of Iraq, to the Shoulder Sleeve Insignia for War-Time Service (the combat patch) ceremony in January 2007, to our monthly organizational days, all the way through to mentoring sessions by MG Joseph F. Fil Jr., the 1st Cavalry Division's Commander and CSM Neil L. Ciotola, the Senior Enlisted Leader for III Corps and Fort Hood in the rear, and during this deployment as the Senior Enlisted Leader for Multi-National Corps – Iraq (MNC-I), Purple Heart and Combat Action Badge award ceremonies, and innumerable promotion, re-enlistment and end of tour awards ceremonies; we attended as many as humanly possible.

CSM Livengood's NCO induction ceremonies were really the most memorable and impactful of all the events and activities we led. The NCO induction ceremony serves as a rite of passage for newly promoted NCOs. The Detachment Sergeants and First Sergeants supported CSM Livengood in building and developing a cohesive bond and esprit de corps during these professionally scripted and well-rehearsed events. Although I was not an automatic attendee to the NCO induction ceremonies, my CSM invited me and I accepted. I was grateful for the invitation to witness the induction ceremonies and to hear the guest speakers impart motivational words of wisdom based on their proven experience. My favorite portion of the NCO induction ceremonies was when all of the NCOs in attendance recited the NCO Creed with vigor and enthusiasm. To me, this ceremony embodies the foundation of the Army and the fact that we are always concerned about our people and our mission. The foundation of the Army is built on trust between the leader and the led, amongst peers, and between the military, and the American people. We are simply expected to be tactically and technically competent while remaining humble, yet confident. The fact is, we as an Army are successful because we have strong NCOs.

We had a certain flow to our events that seemed to work for Task Force Bronco.

The goal for most of these sessions was to break up the routine and allow our Soldiers to interact together, even if just for a few moments. So we tried to mix it up. For one of these events, I asked every single officer in the battalion to write an article titled, "Why I Serve." The battalion S3 took the lead and submitted our young officers' articles to '1775', the journal of the Adjutant General's Corps Regimental Association. Twenty six Task Force Bronco troopers had articles published in the Summer 2007 edition.

But, with all of these events, with everything we planned, and everything we had done together, nothing really prepared us for what it would be like to plan a memorial ceremony for one of our own. SSG Clamens' memorial service was by far the most difficult event we ever planned.

Sitting here with you today, I can admit that I was in no shape to plan a memorial. I was overly emotional and still dealing with the loss of my Soldier and both the seen and unseen injuries of our other Soldiers. No matter what I did, I couldn't seem to stop the tears, the flashbacks, and the overwhelming sense of helplessness that quite literally swamped me whenever I thought about the lives our young Soldiers left behind.

Lives I was supposed to return them to.

That day, when I walked into the Camp Liberty chapel during the memorial rehearsal, my strength came from knowing my team had every detail taken care of. I didn't have to worry about a single thing, because the Battalion's XO, S3, and CSM had the event planned right down to the second. I trusted them to maintain a spirit of excellence.

One of my sister battalion commanders, LTC Darrell Duckworth, who had the call sign of "Stagecoach 6", of the 68th CSSB was a Godsend to me during this time. He actually reached out and offered to loan me his battalion chaplain for the service because we did not have one assigned to our unit. Darrell also offered me wise and much needed counsel, along with a steady listening ear. Although we were peers, in many ways he assumed the role of big brother to me. His unit's size and scope dwarfed mine, yet our responsibilities were very much the same; to provide resources and guidance for our respective subordinate elements so they could complete the myriad of missions we were charged with achieving.

The 68th CSSB provided command and control of medium (petroleum and general cargo) and heavy-lift transportation support, tailorable ammunition support, direct support maintenance, bulk petroleum, and Class IX, or repair supplies for maintenance, to divisional and non-divisional units in his area of operations. LTC Duckworth's battalion tracked flights for Soldiers and performed perimeter guard, executed transportation movement requests, conducted combat logistics patrols. His Soldiers were constantly driving on the dangerous roads to transport millions of gallons of Class III substances, including petroleum, oils, and lubricants, as well as other commodities. His Soldiers were the ones my Soldiers graciously provided personnel services support to.

In fact, it was LTC Duckworth who informed me that the Chaplain Corps actually takes the lead on memorial services during deployments and in garrison environments.

For us, it was a responsibility we were happy to place in their capable hands, as we were all still struggling with the loss of our Soldier. Walking through our grief made it difficult to think about protocol and correct order of a specific ceremony. Using this approach meant we did not have to create our own template for a proper and dignified way to honor SSG Clamens. Between the 1st Cavalry Division's chaplain's office, the 15th Sustainment Brigade's staff, and the 68th CSSB's leadership, Task Force Bronco was well supported.

The memorial service seemed to fly by and crawl all at the same time. I remember all of the movements being so crisp and precise. The order of events was very respectful and well thought out. Everything started when the chaplain led us all in a nondenominational prayer followed by the playing of our National Anthem. Master Sergeant (MSG) Edwards read the blessings listed by Jesus in the Sermon on the Mount found in Matthew 5:3-11 during the templated scripture readings. I spoke, CPT Graves spoke, and then during the portion where someone who personally knew the Soldier well speaks; CPT Cristian Pinzon, the Battalion S1, stood up and spoke with an indescribable reverence and strength.

CPT Pinzon was able to share stories of how SSG Clamens conducted herself throughout our 12-month deployment in Iraq. He mentioned her "can do" proactive, positive style of interacting within the Camp Adder FOB. SSG Clamens was a 35-year-old NCO assigned to the 834th Adjutant General Company (Postal) (Direct Support), a U. S. Army Reserves unit located in Miami, Florida. Her military occupation specialty was an Administrative Specialist where she worked directly for the 1SG and Company Commander during this deployment as the unit's company clerk. She was a professional NCO, who took a great deal of pride in submitting the myriad of routine personnel status reports, as well as the personnel actions for the members of the 834th. She advocated for her unit and was admired and respected for her technical competence and her big 'ole smile. SSG Clamens would often help out in the mail warehouse where the volume of mail was always high. She was a married mother of three young kids and was affectionately known as "Lilly."

Then, Private (PV2) Behrens shared a slideshow she created filled with pictures overlaid with beautiful music. This was followed by the chaplain's eulogy and then the moment we all met with equal parts dread and anticipation, the roll call, the playing of TAPS, and the 21-Gun Salute.

As a Soldier these moments rock you to your very core. We've all heard the saying 'all gave some and some gave all.' It's something that people tend to roll off the tongue without ever really stopping to think about what it actually means. Yet, in that moment, when you hear a Soldier's name called for a roll call they will never answer, the meaning hits home with clear, brute force.

While I can attempt to paint the picture, it is truly something you cannot fully comprehend without experiencing it first-hand.

After the chaplain shared a heartfelt eulogy, one of the senior NCOs stood for the roll call, walked slowly to the front of the chapel, and centered herself in front of a group of Soldiers seated on the third row of pews to the left side of the chapel. Then, a full squad of Soldiers rose in unison, standing quietly in the position of attention, silently waiting for the NCO. She began conducting a roll call of each Soldier standing. This ceremony is meant to represent the deceased Soldier's last formation.

Most Soldiers participate in a version of this ceremony each and every time they report for duty. They stand at the position of attention while the Platoon Sergeant calls out each name and they clearly answer, "Here, Sergeant!" During a typical day, each Soldier responds and the squad leaders disseminate additional information to their squad members and then the Soldiers move on to other things.

But a memorial service is obviously not a typical day.

Master Sergeant Patrice Moses called out each name and received a response with each one until she called, "Staff Sergeant Clamens!"

Obviously, there was no answer to her call. So, she waited a moment or two and tried again.

"Staff Sergeant Lillian Clamens!" Nothing.

Once more the senior NCO called her name, "Staff Sergeant Lillian Ladae Clamens!"

Her voice rang clearly through the room, yet her call still went unanswered. Until three volleys of shots cracked through the night and brought us collectively to our knees. As the weight and significance of the moment set in, the call that went unanswered was followed by twenty-four of the most powerful musical notes ever strung together as TAPS began to play.

One of our teammates was gone, and there was nothing we could do but grieve.

As the sound of each musical note flooded the chapel, a new wave of sobs racked through my body. And it was not just me. From around the room I could hear others fighting through their own pain and I can promise you there was no shame. This precious woman paid the ultimate price and we were all hurting from our loss.

It is a moment, a ceremony, which I will never forget. Even then, the number of people who chose to attend to show support for our battalion struck me. The general officers, brigade, and other battalion leadership were all in attendance in an effort to pay their respects to SSG Clamens, while also supporting the rest of us.

As for the Broncos, we were all there. The majority of our battalion was in attendance with the small exception of the 12-hour casualty shift workers, the battalion EOC, and a handful of others.

At the close of the service, each of the command teams lined up to pay their own respects. One by one the teams would step forward to stand directly in front of a symbolic display of her dog tags draped over her upside down rifle with a ballistic helmet on top and a pair of combat boots placed directly in front of a handmade memorial plaque designed by SSG Graves. Each command team or pair of Soldiers in attendance saluted in unison and dropped their salute in a slow count. Then, one or both would leave a unit coin or combat patch on the table and depart. The next command team would move forward in the exact same fashion.

I had attended other memorial services in theater, however, for me this was the first time that I truly felt the void and understood the importance. It was the first time I felt the raw hole in my stomach grow as the Soldier missed her roll call and TAPs was played. Not only that, but as I sat there I thought about and worried over the impact to my unit as a whole. How could they possibly be feeling as they attended a memorial service for a Soldier who just days before had sat beside them joking and laughing as she planned to finally travel home?

SSG Clamens was much more than just another Soldier to me. She had a smile that lit up any room she was in. Every time I talked with her we would speak about the overall postal and personnel service support mission, as well as her kids and my son. That is what I remember most. Every single person who interacted with SSG Clamens knew, without a doubt, that she was a proud mom. She, much like me, would constantly interject her kids into her conversations when she was not focused on performing her daily mission and when it was appropriate. She had established close bonds with Soldiers within her unit and with other NCOs on the FOB. They played cards together, ate dinner together, and attempted to maintain some semblance of normalcy in a deployed environment.

She was not, and is not, just another Soldier. She is mine. And she is missed.

After the service ended, LT Capetillo carried the task of removing SSG Clamens' gear from the locked area and inventorying her personal effects. These were sent, along with the appropriate items from the memorial service, to her family back in the States.

CHAPTER 8
A New Nightmare

"Each one that has worn this uniform before me is a true hero and should be treated as such with high respect and honor for their service to this wonderful nation we live in."

- SFC John J. Tobiason

(Excerpt from an article published in Hayfield, Minnesota's Star Herald in December 2007)

When you are deployed you tend to count the days. Not only the number of days you are in theater, but also how many you have remaining until you can finally begin your journey home. Ask any deployed Soldier and they can tell you, almost without thinking, how much time they have left.

At this point for Task Force Bronco, it was 44 days and a wake up. We had less than two months remaining in Iraq when I found myself sitting in a military courtroom in Balad, Iraq answering pointed questions under oath. Not something anyone ever looks forward to doing, but this particular event was harder than most.

Why?

Because 49 days after the absolute worst day of my military career, (and easily one of the top-six worst days of my life) I was required to be present as SSG Clamens' commander was undergoing court martial proceedings. It was a serious occasion. And though the charges were completely unrelated to her death, participating in this process brought the grief back all over again, but it was merely another aspect of serving as a battalion commander. I remember sharing my outrage and utter disbelief when SSG Hilliard, who was recovering and undergoing continued treatment on her amputated leg, informed me that she was being told she was required to return to the Iraqi theater to participate in this court martial proceeding. She was trying to find a way to return to the theater in the midst of caring for herself. Needless to say, I ensured that the military lawyers received whatever pertinent information they required of her without her having to be physically present.

Once the proceedings were over, I left as soon as possible and flew back to Baghdad. The Captain would end up serving some months behind bars as a result of his actions during our deployment, but my part of the process had ended.

Forty-nine days after a barrage of mortars rained down on Camp Liberty, Iraq, I was blessed to live to see my 41st birthdate, which I share with my Operations OIC, the Battalion S3, Captain Jeannie Deakyne. Captain Deakyne was married to another Captain who was also deployed in the Baghdad area of operations, and she represented many of the Task Force Bronco heroes who had diverse experiences in and out of a military uniform. CPT Deakyne, who had served as a Brigade Personnel Officer (S1) for the 4th Brigade Combat Team, 1st Cavalry Division, and as the Commander of Charlie Detachment, 15th PSB during OIF II was a great asset to our team. She initiated the application process for the Mrs. Texas International pageant, while deployed on Camp Taji, Iraq during OIF II and subsequently won the competition in May 2005. Task Force Bronco was fortunate to have her serve as our Battalion S3 during OIF 06-08.

The 15th PSB's S3 was the staff element responsible for routine operations of the battalion's processes and procedures. They managed internal training programs as well as ensuring the battalion's personnel military education was ongoing. A typical battalion S3 holds the rank of Major or Lieutenant Colonel. However, CPT Deakyne was uniquely qualified to serve as the Task Force Bronco S3. She had previous combat experience in the Iraqi theater and had served in one of the newly transformed Brigade Combat Teams and as a Detachment Commander within our unit as a subordinate commander. She was married to another officer and understood what about 50% of our task force members were experiencing when balancing careers and families. She also had the unique perspective of being married to an officer who was constantly participating in road convoys. I leaned on her a great deal for many major and minor tasks. She was the type of officer who was committed to service with the warfighter at the forefront of her decision paradigm.

While serving as the Task Force Bronco S3, CPT Deakyne oversaw the Battalion Emergency Operations Center (EOC), which was the hub of the entire task force. She and her team of forward thinking, proactive, leaders skillfully managed and processed the myriad of tasks, directives, and guidance which came from our personnel technical chain within the human resources community and our operational chains of command which began with the 15th Sustainment Brigade, located in Taji. The S3 staff planned and ensured the execution of the unit's daily operations, and planned future operations through the inactivation that was looming when we redeployed to Fort Hood, Texas.

CPT Deakyne's mother and family members sent her care packages on a nearly weekly basis. The packages included basic starter kits for every itty bitty national celebration. She took a great deal of time and effort to ensure that the operations section, which was the largest on the battalion headquarters staff, took the time to enjoy the experience of leading the personnel support mission within the theater.

Because we had junior NCOs and Soldiers assigned to the 15th PSB working on the headquarters staff in the Intelligence (S2) and

Communications (S6) Sections, CPT Deakyne, the battalion S3 and her NCOIC, MSG Nyeedra Edwards, maintained oversight of the these battalion staff elements and fondly referred to the collage of administrative sections as the S11 (which did not formally exist). The battalion S2 was responsible for collecting and analyzing intelligence information about what the enemy was doing, or might do in our area of operations. The information these young NCOs, under the direction of CPT Deakyne and MSG Edwards, provided was priceless as they ensured that the CSM and I were always aware of the significant dangers our outlying subordinate elements were facing. The battalion S6 directed all associated communications efforts. We were grateful that the 1st Cavalry Division's communications office was very familiar with our communications capabilities. MG Fil's G6 took personal responsibility for ensuring that our young signal NCOs were trained and prepared for CSM Livengood's and my insatiable appetite for every electronic device that was available to every other unit in the theater. They coordinated to have computers, fax and copiers, phone systems, and both secure and unsecure modes of communications available to our higher headquarters and to our family support group located in the rear at Fort Hood, Texas. They coordinated live VTCs with our Family Support Groups in the rear, which offered a tremendous boost in morale.

CPT Deakyne and I, as well as the other members of Task Force Bronco, were physically, mentally and emotionally drained by this time in our tumultuous tour of duty.

When I flew back to Camp Liberty after the court martial was over, and finally arrived at Pad 11 that afternoon, I was looking forward to a somewhat dull and uneventful evening. But first I had to check in at battalion headquarters.

Our work space was always bustling with activity as people engaged in exciting topics of conversation or worked swiftly to complete tasks for our higher headquarters, or in support of one of our geographically dispersed subordinate elements. But, as soon as I walked through the door of our battalion headquarters that day I knew something was wrong. There were only a few people milling about and no one would make eye contact with me.

Despite the battalion headquarters being festively decorated, I noticed that most of the lights were turned off at the end of the hallway. When I got to my office door, I noticed balloons, colorful party favors and small, multi-colored decorations on my desk. My first thought was that someone was playing a joke; I honestly kept expecting the lights to come on and hear everyone yell "Happy Birthday!"

I realized something significant was going on and stopped on the spot, asking that someone give me a status report.

MSG Nyeedra Edwards, approached me with a solemn look on her face. She informed me that a serious incident had occurred within the past 30 minutes and that we needed to go to the TMC as soon as possible.

"Ma'am," she said, still struggling to look me in the eye. "Something just happened. It's Sergeant Tobiason. He's been shot."

I verified which clinic and quickly took off across the base toward the Camp Victory TMC. As I drove to the clinic I felt that all-too-familiar sick feeling settle in the pit of my stomach. This was the same clinic where CSM Livengood and I sobbed together only a few weeks prior. And I felt the exact same helplessness all over again. As I walked through the door of the clinic I called out to whoever might be within earshot. I wanted to see my Soldier. I remember thinking that maybe they got it wrong.

Someone, I was in such a daze I honestly cannot remember who it might have been or anything about them, led me to what looked like the exact same long, thick, black bag with the exact same overly noisy zipper. Although I didn't want to, I had to see for myself. I had to be sure so I asked that they show me. You could almost feel the sorrow swelling in the air as they lowered the zipper.

It was him. It was my Soldier. Damn.

It was John J. Tobiason, recently promoted to Sergeant First Class. He was a breath of fresh air. This guy was always upbeat and willing to take on any mission.

SFC Tobiason deployed with the Detachment 4, 847th Personnel Services Battalion in October 2006 as part of Task Force Bronco. During the early phase of his deployment, SFC Tobiason served as a member of our Battalion's EOC at Camp Liberty, reinforcing indirect fire drills to the majority of the eight detachments and company-level subordinate elements in theater. CSM Livengood, then relocated him from Camp Liberty to Camp Striker where he served as a liaison officer (LNO) at Striker Stables.

In this role, he was responsible for updating all transient passengers, located at the Striker Stables, on helicopter and fixed-wing flight information. He remotely manifested passengers for outgoing flights to FOBs in the Baghdad Area of Operations via the BIAP. He also provided hourly updates for all scheduled flights in and out of Baghdad International Airport (BIAP). He served short stints as a shift leader and as a platoon sergeant at BIAP in support of our R5 mission: Reception, Return To Duty (RTD), Rest & Recuperation (R&R), Replacement, and Redeployment. SFC Tobiason was promoted during this rotation on 10 May 2007 and asked me to reenlist him when he became eligible in December 2007.

What is significant about this is that SFC Tobiason's unit was scheduled to return to their home station in early October 2007. But he didn't. Instead he, along with four other members of his unit, chose to voluntarily extend their deployment through January of 2008. This meant he would wait and rotate home with the rest of Task Force Bronco instead of taking the opportunity to go home with the rest of his U. S. Army Reserve unit.

His loss hurt deeply.

In truth, the loss of another Soldier, just 49 days after the mortar attack that killed SSG Clamens, injured SSG Hilliard, and rattled so many of us, was both humbling and devastating.

Hearing about the details of the incident didn't make it any easier.

On 28 November 2007, SFC Tobiason and another NCO ventured to the other side of our FOB to watch the Washington Redskins' cheerleaders perform as part of the USO tour. It appears that as they stood and watched the show, as so many others did, SFC Tobiason was shot as a result of an apparent stray round. Because of the location of the event, several medics were in the area and began rendering aid immediately, but to no avail; SFC Tobiason was already gone.

In that moment, I knew I would not be alone in my bereavement, shock, and dismay. SFC Tobiason was loved and there would be many, many people actively grieving his loss including his faith-filled family, friends, and loved ones.

Although SFC Tobiason's original unit, Detachment 4, 847th Personnel Services Battalion, had already demobilized, they would be made aware of this tragedy.

I often reflect back on the first day I met SFC Tobiason early in November 2006, one year prior to his death. CSM Livengood escorted him into my office as we were still determining the skill sets needed within the battalion headquarters to support our grand mission. SFC Tobiason was eager to serve and willing to support the unit. After the initial meeting, the Battalion Leadership Team decided he would best serve in the battalion operations cell.

A few months later, SFC Tobiason's company commander hosted his promotion ceremony and he was promoted to the rank of Sergeant First Class. I remember how happy he was and how, after the formation was released, no one rushed away. Each member of his unit and many members of the battalion's headquarters element stood in line to shake his hand, share a congratulatory remark, and acknowledge the hard work he obviously expended to earn the distinction of becoming a senior NCO in the greatest Army on the planet.

As his unit moved closer to rotating back home, my CSM informed me that five members of the Detachment 4, 847th Personnel Services Battalion had requested to extend beyond their original 12-month deployment. Administratively, it was a Lieutenant Colonel-level signature required to process the extension request, meaning they needed me specifically to sign off on it.

At first the entire request seemed odd to me. I had countless active duty Soldiers who were praying to go home early and could not due to the troop surge and mandatory 90-day extension of our original 12-month deployment. Because it didn't make sense to me; I told the CSM that I wanted to talk to these Soldiers. I needed to understand why they were making this request before I could, in good conscience, agree to it. Above all the others, I clearly remember my conversation with SFC Tobiason.

He shared with me how strongly he felt about being a member of Task Force Bronco. He couldn't imagine leaving all of us in theater while he went home. Instead he wanted to return home to Minnesota only when the rest of us left for Fort Hood, Texas. To him the mission was not over and he didn't want to leave us to finish it without him. I signed the extension request.

A few weeks later SFC Tobiason knocked on the door of my office. He asked if I would be willing to reenlist him and I gladly said yes. The reenlistment process requires a commissioned officer to render the oath of office and sign the enlistment documents. The fact that he asked me was a huge honor because there were plenty of officers he could have asked. But then he kept asking.

At least once a week he would stop by my office and ask if I was still going to reenlist him. The first time I was happy to reassure him, saying that I was looking forward to participating in his reenlistment ceremony. But after the second, third and even fourth times, I began to wonder if there was some kind of joke that I didn't know about. I asked my CSM if there was something I was missing, and he reminded me how excited SFC Tobiason was to be part of our unit. To him, having the BC host his reenlistment ceremony was a huge deal and he couldn't wait for that day to come.

CSM Livengood's perspective was invaluable and from that day on I began looking forward to SFC Tobiason's random stops by my office. Our interactions were short, only about 2 minutes each, but they always served to brighten my day.

Now, just a few days after his most recent visit to my office, he was gone.

That is the reality of life is a war zone. You're at war. Anything can happen. And sometimes it hits far too close to home.

SFC Tobiason was a dedicated, enthusiastic Soldier who was proud to wear his uniform each and every day. I am honored to have served with him and only wish his story had a different ending.

Shortly after his death, I had the opportunity to speak with SFC Tobiason's mother briefly. I must say that it was very evident where he got his positive perspective on life. His mother, bless her, through her sobs expressed to me the wonderful support that she and his sister and brother received from the military. She told me he loved the Army and was proud to serve his country and that knowledge somehow seemed to provide comfort to her.

She told me that she was happy he died doing what he loved. It was difficult to hear because I could not imagine having to say those words myself. As we spoke, I told her we would have a memorial in his honor and asked if there was anything she wanted me to share.

SFC Tobiason's sister wanted me to relay that he felt that the military was just as much a part of his family as they were and that he was very proud of his service and proud of what he was doing. He had a passion for the military. His sweet mother told me that they are a family of strong faith and know that God has a plan for our lives. She asked if he was alone and I was thankful to be able to tell her no, that one of our other Soldiers was with him when he died. She thanked me for calling. I thanked her for her time.

I can remember sitting at my desk and reading a letter SFC Tobiason wrote to the newspaper in his hometown, Hayfield, Minnesota's Star Herald. His words were so powerful and painted such a clear picture of not only his thoughts, but his heart. He wrote this letter in 2007 during his deployment, not long after his promotion ceremony. With their permission I would like to share his words with you now.

Dear Editor:

Hello people of Hayfield. My name is John Joseph Tobiason. I just recently got promoted to Sergeant First Class from Staff Sergeant. I am currently stationed here in Iraq with the Army Reserve unit I belong to out of Fort Snelling, MN. I volunteered for this deployment and have been enjoying every minute of my tour here so far. I will be extending here within the next few months to stay here longer to meet re-enlistment specifications. (Tax-fee bonus) I'm proud of what I do and am also a proud member of Post 330 there in Hayfield. I take being a veteran serious because it's my way of saying thank you to those who gave us the right to have and enjoy the freedoms that we have now. Each one that has worn this uniform before me is a true hero and should be treated as such with high respect and honor for their service to this wonderful nation we live in. Many people have died to give us these rights and should always be remembered for their accomplishments. Those are the true heroes of the world. I would never change my mind about why I do this for my job and career. I love my country and I know that if I was to die for it, it would've been an honor dying for something so precious as freedom is for all of us. I know a lot of you in town know me and some don't, but I will be looking forward to getting back to my home there on First Avenue as soon as I'm ready to leave here. I wish all the best for everyone in the town. Take care and may God bless each and everyone in the town.

SFC John J. Tobiason

Iraq

CHAPTER 9
Finding Home

"Some succeed because they are destined to; most succeed because they are determined to."

– Unknown

After seeing SFC Tobiason, who was easily the most upbeat Soldier in my unit, lying in a body bag, I knew it was time to act. I was going to control what I could control and start moving my troops back to the States.

I left the TMC, returned to the battalion headquarters and walked directly into my battalion S3's office. MSG Edwards was in the office alone and as soon as I shut the door behind me she knew something was up. I told her point blank to get the Soldiers out of the theater as the mission was reduced. Get them home; get them gone as soon as possible.

I may have used a few curse words.

My thought was that only mission-essential personnel needed to remain in Iraq. I was not willing to wait for our higher command to give us a date to pull our Soldiers out of the country. Instead, I intended to be proactive and present the Corps G1, which handles personnel, including readiness, with a brigade S1 certification plan, which would prove we had trained our replacements properly and would render our services no longer necessary in the theater.

In the meantime, our chain of command changed. The 1st Sustainment Brigade out of Fort Riley, Kansas, assumed responsibility for the MND-B and MND-C logistics mission from the 15th Sustainment Brigade. My boss and his organic battalions redeployed to Fort Hood after providing outstanding logistical support, leadership, and guidance. The 4th Infantry Division, led by my former college ROTC instructor, MG Jeffrey W. Hammond, replaced the 1st Cavalry Division as the lead for MND-B and moved into the Division Headquarters on Camp Liberty. The 3rd Infantry Division assumed responsibility for MND-C and occupied space on Camp Victory. A second division headquarters full of people were living and eating and requiring personnel service support in the November and December 2007 timeframe. Victory Base Complex was bursting at the seams and bustling with activity. Five additional Brigade Combat Teams were in the Iraqi theater with the majority of them residing on Victory Base Complex and in and around Baghdad proper.

President Bush's troop surge increased the total number of U. S. Army brigades in country from 15 to 20, or an increase of just under 20,000 Soldiers. The U. S. Marines and a myriad of support units and contractors were either extended or had recently arrived in theater. This was the playing field and it was shifting daily. My goal was to stay on top of these changes so I could still move my troops out of theater as soon as possible. I was focused on how to reduce Task Force Bronco's footprint within the context of the troop surge.

When the 1st Cavalry Division and the 15th Sustainment Brigade returned to Fort Hood, our mission and mindset changed. We focused on training and certifying Brigade level Human Resources (HR) sections to conduct Essential Personnel Services, as well as transitioning the military postal mission to contractors. The casualty tracking mission and personnel accountability mission remained largely unchanged. We quickly reduced our footprint whenever and wherever possible. I sent the XO back to set the conditions for our redeployment and inactivation. My intent was for as many Soldiers as possible to return to the States prior to the end of December so that they could spend the Christmas holiday with their families and loved ones.

My impression of my new Brigade Commander was always positive. He was very inclusive and supportive. He consistently invited CSM Livengood and me to his functions and he attended our organization day, SFC Tobiason's memorial service, and also made a point to visit with Soldiers throughout his area of responsibility. I was, however, put off by his staff.

I am sure it was my issue after having been in theater for more than 13 months and them having just arrived. Our interactions mostly revolved around how we were not doing the R5 and postal mission like they expected. My tolerance for being second guessed was low. I continued to invest my time and energy towards the members of Task Force Bronco, especially as we were approaching the holiday season in the midst of this dramatic increase in personnel. We began transferring several aspects of the theater's postal mission to civilian contracted organizations. Interestingly, the postal contractors arrived undertrained and understaffed, which required our Soldiers to remain in the post offices longer than expected.

The personnel community was beginning to shift to an HR paradigm under the Personnel Services Delivery Redesign (PSDR). The AG community's answer to the Army's overall transformation from a division–centric to a brigade-centric management was in full effect approximately one year after implementation in 2006. PSDR was touted as a way to reduce slow, inefficient, ineffective layers of personnel processing. The Army was close to inactivating all of its Personnel Services Battalions in all components. The structure was moved into most of the brigade-level HR sections and facilitated the capability of providing EPS and personnel strength accounting. The R5, casualty and postal missions were moved to newly established HR companies, led by an AG major. Multiple layers of a newly established

HR hierarchy were created to provide technical oversight and guidance to this new phenomenon.

From my vantage point, it briefed well, however the actual execution was awkward and undermanned. Leaders from all sectors were attempting to get up to speed, while continuing to provide support to their own Soldiers. I knew, however, like everything else in the Army, we would figure it out. We had to test the new paradigm and ensure we continued to support the warfighters.

The new structure would surely be tested during the next rotation with the 1st, 3d, and 7th Sustainment Brigades' arrival in theater in support of OIF 07-09. One of the challenges for Task Force Bronco and the 1st Sustainment Brigade was that almost all of the 510th Human Resources Company's subordinate elements had deployed on their original schedule, meaning they arrived in Iraq months before their command and control element. I assumed it was an oversight due to the extension of active duty elements while the reserve component elements remained on the same deployment timeline. Some of the smaller active duty elements apparently remained on their same deployment timeline as well. Because some of the brigade combat teams were deploying into the Iraqi theater in various stages of PSDR and the 510th HR Company was not arriving in theater until the end of December 2007, we prepared for a TOA in early January 2008. The fact that the 15th PSB had assumed command and control over the smaller HR elements only added confusion to an already confusing situation. It was an interesting mix of legacy and modular formations throughout our area of responsibility.

Beginning in the August 2007 timeframe, we started tracking smaller and smaller sized elements. On a few occasions we found out that we had 3 personnel in Kuwait without a command and control element. The casualty platoon consisted of a platoon leader (a LT) and platoon sergeant (a SSG or a SFC) who were intended to be able to lead and train platoon members from any location. This was an interesting concept indeed, because this was some of the same rationale used to determine why the legacy PSBs should inactivate; no established relationship with the brigade or division they would support during a deployment or field training exercise.

The casualty platoon leader and platoon sergeant had to quickly assess the level of experience of the people they were meeting in theater. So, this required CPT Graves, the HHD commander, who I gave the additional duty of Unit Movement Officer, to track all of the inbound and outbound subordinate elements. He had to process, track, and engage with six different unit identification codes, just like a plane ticket is tracked when an individual flies from one location to the next. These 3 and 4 person elements deployed on different timelines. They were a mixture of active and reserve component Soldiers with and without previous combat experience. The good news is that we received, housed, and trained them all. The not so good news is that we received, housed, and trained them all.

During our training and certification phase, we quickly realized that one of the challenges was to identify an HR element that would provide personnel service support to Soldiers, civilians, and contractors who were not assigned or attached to one of the newly formed PSDR Brigade Combat Teams. The issue was location. We had to determine who would support outlying units that were not assigned or attached to a centrally located element in a forward deployed theater of operations.

PSDR mans and equips at the Brigade and Battalion S1 level, but does not provide for the outlying company or detachment located at a Patrol Base or Strong Point, which are essentially smaller FOBs without fortifications. Most of this is mitigated (in theory) by the internet, using digital senders and e-mail as a means of transmitting evaluations, promotions, and actions between geographically dispersed units and their S1s. The fallacies that prevent this assumption from solving the issue, however, are that connectivity is unreliable, scanners and digital senders break down, and when you're a Soldier conducting patrols day-in and day-out, providing your own

PSS is not the priority of effort. Additionally, certain elements of PSS, to include CAC production and CAC PIN resets, require the HR Soldier to be physically present for the service.

One of most impressive outcomes of looking at a task from a fresh perspective was how the 1st Sustainment Brigade's Support Operations Section (SPO) revamped the postal distribution plan within the MND-B and MND-C battle space. The bottom line was that their plan ensured the mail was delivered in a timely manner and actually decreased the overall average time before Soldiers' received their letter and packages from their loved ones. I am not sure of what it is currently; however, during OIF 06-08 I am pretty sure the military postal standard was 12-18 days. During the 2007 holiday season it was 10 to 12 days. Some Soldiers reportedly received their packages in 5 to 6 days, which is pretty amazing when you consider all of the factors involved.

And then, after all of this seemingly endless activity, we were done. The day had come and it was time for us to finally go home.

On 2 January 2008, the 15th Personnel Services Battalion cased their colors during a transfer of authority ceremony in Baghdad, Iraq with the 510th Human Resources Company based out of Fort Eustis, Virginia. Thus marking the end of Task Force Bronco's service in Iraq.

The four Soldiers who had extended along with SFC Tobiason headed home. A small contingent of the 15th Personnel Services Battalion and I departed BIAP enroute to Fort Hood, Texas. Our mission was over. It was time for the Army to truly test the PSDR concept and we wished them well.

As for us, we were going home.

I was 39 years old when I first assumed responsibility for this intriguing mission in Iraq and 41 years old on 2 January 2008 when we transferred our authority to the 510th HR Company. It was a long road no matter how you walked it. However, the longest time seemed to occur during the three days it took me to get to my son after leaving Baghdad and finally arriving at the welcome home ceremony on 5 January 2008.

The trip home after a deployment seems endless. We were traveling for several days; sleep deprived, hurrying to wait for the next announced milestone on our travel itinerary, and longing for a long, hot bath. I talked with my mom about not coming as I had done previously when I was a LT and redeployed from Saudi Arabia in 1991 and as a Major when I redeployed from Qatar in 2003. I told her that of all of the things the Army excels at, they are not great at returning us to our home stations on the date and time that is first announced. The flight times always change and family members are alerted within hours that their loved ones were returning home.

My Mom didn't care a lick. She was coming no matter what.

She came with my niece, my son and one of my closest friends. There aren't even words. I want to try and tell you what it feels like. The mix of excitement and joy blending with small amounts of uncertainty that threaten to overwhelm you on your way home.

Mostly I was worried about my son. I had done everything in my power to keep my voice in his sweet two-year-old ear throughout my entire deployment. I threw an over-the-top, superhero-themed birthday party for him during my scheduled R&R trip to Mississippi in June 2007. I dressed him in a superman costume and I dressed in a fabulous superwoman costume. Throughout our separation, I recorded videos of me reading books, I talked to him via webcam, and I called as often as I could. But I still worried that he wouldn't remember me. I had left a baby and I was returning home to a toddler.

There were so many emotions.

When we finally landed in West Fort Hood, I knew the only thing standing between me and my boy was turning in my weapon, a reintegration briefing, a 15-minute bus ride and the division commander's speech. None of that could happen soon enough for me. It was like everything was moving at a snail's pace and I was stuck at the back of the line.

The buses pulled up to Cooper Field, which was the length of a football field or better. We saw a crowd of people, some with signs, cameras, new clothes, and balloons. I remember how the weather fit the occasion; it was a gorgeous day.

SSG Hilliard met us as we stepped off the bus. Seeing her brought a sense of joy to my heart. I noticed she did not have her prosthetic yet, so one of the NCOs pushed her in a wheel chair as we marched in unison in a mass military formation across the parade field. MG Fil, the 1st Cavalry Division's Commander, gave a brief and inspirational speech, and then everyone in the formation moved around the field to find their loved ones. It was time.

Finally!

My niece spotted me first.

She was moving towards me, carrying my two-year in her arms. I stopped in midstride and allowed him time to recognize me. He tilted his head to the right and said, "That's Mama! Mama is out of the box." My niece described the scene as "love wrapped all around." All I know is the feeling of seeing him, of wrapping my arms around his tiny body filled my heart like nothing else could. With my son in my arms, I knew I was finally home.

I made my way to my mother, who was seated in her wheel chair with tears rolling down her cheek. I introduced SSG Hilliard to my mother who was also an amputee. My mother remarked to SSG Hilliard, "Baby it's a lot you can do with one leg, I know because I have one leg." I introduced my mother to CSM Livengood, who she often referred to as my bodyguard. I could not get her to understand that he was my battle buddy. My Brigade Commander and the Division Commander also met my mother and son. The Division commander placed his highly coveted Stetson on my son's tiny head. I knew I mattered to him and I knew he valued the 15th PSB.

I remember watching SSG Hilliard. Several Soldiers took pictures with her, she possessed such poise and exuded a sense of resilience I had prayed for her and her family over the previous months.

I, along with all of the recently returned troops, signed out on a 3-day pass. I think I slept for a day and a half in the hotel. Each time I opened my eyes, my mother would be sitting in one of the chairs in the hotel watching me sleep. I tried to engage with my son, but the time zone change, and the lack of sleep proved to be too much.

I did not have a place to live or a car to drive. I had only been on Fort Hood for about two weeks prior to deploying in October 2006. I had sold my minivan prior to deploying. I had rented a room on post until my on post house was to be available in February.

When I finally woke up. I gave my niece my credit card, told her to catch the first flight available because I wanted to immediately start bonding with my boy. My mother refused to leave. My youngest brother packed up all of my son's belongings and transported them from Mississippi to Texas, helped me set up his room, picked up my mother and drove back to Mississippi.

I returned from my 3-day pass to begin the ten-day reintegration schedule, which is mandatory for everyone who redeploys. After being away from the "normal" routines and creating new ones in Iraq, Soldiers must decompress in a safe and healthy environment. Finally, we go on what is termed, block leave; this is where everyone takes up to 30 days of leave. I took about 15 days.

The majority of the Soldiers assigned to the organic unit were already reassigned to a military school, had departed the installation, or were reassigned to another unit on Fort Hood. Our rear detachment and family support group leaders had done a marvelous job throughout the entire 15 months, managing both small and large tasks. The S4 and the S1 sections were very busy turning in equipment and processing awards, evaluations, and coordinating for our Soldier's next duty assignments. The XO had it all under control.

We conducted a flawless final military ceremony to inactivate the storied 15th Personnel Services Battalion effective 15 April 2008.

SSG Hilliard, who was just about to be promoted to SFC, participated in our inactivation ceremony. I do not remember the speeches that either BG Brooks, now a four-star General, or I gave. I do, however, remember the 1st Cavalry Division's band sounding marvelous. The thunderous 1st Cavalry Division's Horse Cavalry Detachment conducting the reenactment of the last charge with drawn sabers during the ceremony was one of the most memorable events of the entire day.

It goes without saying; I will always remember the members of Task Force Bronco.

All gave some and some gave all.

Bronco Strong! – Wagon Masters! – First Team! – Army Strong!

This is BRONCO SIX signing off of the net

EPILOGUE

A Story Doesn't End...It Changes

"It doesn't get easier, you become stronger"

- Unknown

The thing with stories like this one, with such life-altering journeys, is that they never really end. They just change. And, I find that even though we've covered the time I wanted to share with you there is still more to tell. This story is never truly over.

Tim Livengood, my Command Sergeant Major, my battle buddy, was a true American Hero. Tim was inclusive in his mentorship engagements. He mentored and trained males and females, blacks and whites, and any other demographic you can imagine. He simply wanted everyone to demonstrate their effort, care, and compassion for the mission, the unit and especially our Soldiers. He was technically and tactically competent and incredibly proficient. Even though I outranked him and everyone in our unit, and I was overall responsible for our every action and inaction, Tim often mentored me. He was the only person in the entire world who has shared the vantage point of loving my Soldiers as hard as I do. My time with Tim was short. However, it represented the most significant time in my military career. During our time together, we served as the last battalion command team for the inactivated 15th Personnel Services Battalion and our mission was tremendous. Thank goodness I had him and thank goodness we had awesome subordinate leaders.

A 15-month deployment is a long time to work side-by-side someone for 15 or more hours per day. We didn't have breaks. We intentionally developed camaraderie and respect for one another shortly into the deployment because, in that environment, the gaps between our reputations and our character were easily revealed. Although uncomfortable, I love how he held me accountable. Everyone needs an accountability partner. Tim Livengood taught me that I did not know everything and that good leaders share the spotlight while absorbing all of the blame. He also taught me that I could be strong, authentic, and transparent simultaneously. He actually taught me a bunch of things. The most significant lesson I learned from Tim Livengood was to listen to understand with empathy.

I remember once during the deployment we were attending one of the Brigade Commander's updates in the 15th Sustainment Brigade's headquarters, which was located in Taji, Iraq. I do not recall the topic. I only recall that I felt my CSM kicking my feet under the table. I was giving my boss a lot of push back. The air had left the room and I was about two more back-talking words away from being trounced (figuratively) in front of the seven other battalion command teams. CSM Livengood noticed that the tone I was using was not getting me anywhere with my boss. He finally got me to stop talking by sliding a piece of paper in front of me with the words, "listen more than you talk." Afterwards, he told me, although he understood what I was trying to convey, I was not being heard because I was using the wrong forum to challenge my boss. My boss was and still is a great man, who on that particular day in Iraq, cut the young LTC some slack, thanks to my battle buddy, my CSM, Tim Livengood.

After the 15th PSB's inactivation ceremony on 15 April 2008, CSM Livengood served as the III Corps Special Troops Battalion CSM, where he deployed twice with them. Meanwhile, I was fortunate to serve as a Professor of Military Science. I was finally able to find out how Lieutenants were made. I loved serving at the

University of Tennessee at Martin. The school is located in a small rural town on the Northwest corner of small town USA, where the busiest parts of the town were the Super Walmart and the college.

The former 15th PSB S1 NCOIC, Retired SFC Hilliard, now Henry, learned to continue her bountiful life with her three kids during her initial recovery period. She later earned her Master's Degree, retired from our great Army, and remarried. God blessed her and her new husband with a bouncing baby boy. She was medically evacuated from Camp Liberty, Iraq on 10 October 2007 after the barrage of mortar attacks hit as she and others were entering the military dining facility. She was transported to the 28th CSH, then to Walter Reed Army Medical Center, which is where she realized she lost her left foot. She was transported onward to BAMC at Fort Sam Houston, Texas for further recovery. She is resilient, strong, faithful and fearless.

From the University of Tennessee, I was assigned to a job I felt I was born to do, which was to lead the section responsible for making military assignment decisions for every HR officer in the rank of lieutenant colonel and below. I was able to continue to provide mentorship and guidance to my former lieutenants, warrant officers, captains, and majors who were navigating between stateside tours, field training exercises, and additional deployments to Afghanistan and beyond.

The former 15th PSB S1, MAJ Cristian Pinzon called me one morning in 2009. He told me his military chartered flight had stopped in Bangor, Maine. He was enroute to Afghanistan for another deployment. I tried to remain calm and supportive, because I thought he needed a good listener. After I hung up, I must have emptied every drop of tears my body could produce. For weeks I replayed how I had hugged him and how he grimaced because he had been jarred by the mortar blast on 10 October 2007 and had received other minor physical injuries. I assumed he had invisible emotional scars. I always admired how he continues to focus on his family and his complex military missions.

My battle buddy, Tim Livengood brought me to tears when he and his wife Debbie surprised me by attending my promotion ceremony to the rank of Colonel on 12 April 2011. We had been corresponding over the years through Facebook. He insisted he would not be able to attend my ceremony, but there he was. He retired from active duty on 1 July 2012 after serving over 26 years on active duty. He was assigned to the 15th PSB in 2004 directly from the Sergeant Major Academy, located at Fort Bliss, Texas. Retired CSM Livengood deployed a total of five times while serving on active duty: Desert Shield/ Desert Storm, OIF II, OIF 06-08, and twice with III Corps as the Special Troops Battalion's CSM.

I knew my time on active duty was winding down after my mother passed away in June 2011. She was the only person on the planet I sought validation from. Fortunately, I was selected to attend the U. S. Army War College, which was billed as the "best year of your life." I am grateful I met so many proven and tested leaders who were good and decent people during this 10-month period at Carlisle Barracks, Pennsylvania. After earning a second Master's Degree, I was assigned back to where it all started for me, Fort Jackson, South Carolina to assume the positon of Deputy Commanding Officer for the Soldier Support Institute, the U. S. Army post where all AG privates, NCOs, WOs, LTs, and CPTs receive their technical training. The institute also had several other training missions. I was happy to be in the place where AG doctrine is developed, taught, and disseminated throughout our Army. This is where I could learn to what extent the repurposed structure from the inactivated legacy PSBs to the hierarchies of military HR elements was effective.

While attending a prayer breakfast at about 7am in early February 2015, my cell phone kept buzzing. Thank goodness I had it on vibrate. I had a missed call from a friend of mine who I served with at Fort Bragg, North Carolina and again at the Human Resources Command in Alexandria, Virginia. Because she called three times in a row, back-to-back, I stepped outside of the building to return her phone call.

I was simply not prepared in any way to receive the news. Deidre said, "Angie, you know a lot of people love you." Then she said, "We did not want you to find out on Facebook that Tim Livengood passed away." I heard a loud, uncontrollable roaring; a deeply primal noise. It was as if I was having an out of body experience. I could not process the news she lovingly shared, but I was grateful she cared enough to call. Retired CSM Tim Livengood, my battle buddy died on 3 February 2015 in Kandahar, Afghanistan, while serving as contractor, of heart failure less than a week before his final rotation back to the States. He was making plans to reconnect with his family, loved ones, and friends; he was four days away from returning to Texas to be with his wife and see the house that was built during his year-long deployment in the capacity of a contractor in Afghanistan.

Tim Livengood's passing was more than shocking. I could barely bring myself to breathe upon hearing of his death. The fact that he was so close to returning home to the States was jarring and brought me right back to SSG Lillian L. Clamens and SFC John J. Tobiason.

Within about ten minutes of Deidre's call, one of my former battalion XOs, Jon Althoff, called me. He said, "Ma'am, I just found out." We agreed to keep one another informed of the arrangements. He and I along with about seven other former members of Task Force Bronco flew into Texas to pay our respect to Tim Livengood's family. We counted ourselves amongst his loved ones. Although his earthly body has departed, his legacy lives on through me and countless others.

After his oldest son read a personal letter he had received from Tim to all of us in attendance at the gravesite; I knew I had to share our story. Tim's wife, Debbie told me that they were planning to attend my retirement ceremony in June 2015. He was going to surprise me again. I know I mattered to him. However, I am not sure if he knew how much he mattered and still matters to me.

When my father died in early 2015, I knew I had made the right decision to retire. I loved him for always loving me. I learned a lifetime of lessons from my father. He taught me how to drive, how to present myself in public, and mostly he taught me that family is everything.

I transitioned from the U. S. Army after serving over 26 years on active duty in August 2015, grateful for the amazing impact so many leaders, peers, and subordinates had on me.

I love that my mother and father were proud to say that their baby girl was in the Army. Throughout 26 years, lots of things happened; I have been cursed out using real cuss words, harassed, and demeaned; that is part of the reason I remained in the Army. I thought if someone would attempt to do this to me, then what are they attempting to do to the privates, warrant officers, and the lieutenants. I wanted to empower those around me and share my experiences.

One weird phenomenon is, the Army inactivated the 15th PSB on 15 April 2008 yet there has not been more than a 3 week time period in the past 8 years where I have not received a call, email, text, Facebook message, or seen in person one of my former Soldiers, one of my Broncos. It's about relationships, it's about the good, the bad, and the ugly shared experiences.

That's what I want for each of you.

I want you to love Soldiers. I want your Soldiers, or your subordinates, to respect you. If you are in that role, I want you to love being an AG Operator, an HR professional. I want you to be prepared to lead. Don't be concerned with whether you are liked, be concerned with whether you add value, be concerned and focused on the commander's, or the boss', intent and then meet it, or better yet, exceed it.

It was the honor of a lifetime to serve as battalion commander during combat operations especially with this set of heroes, in particular. I am often amazed when one of my Soldiers will say, "ma'am I know you do not remember me, but I served with you." I always surprise them because I remember each and every one of them, if not by name, then by face, deed, or mannerism. Each of them matters to me.

And they always will.

BRONCO STRONG!

A SOLDIER'S WORDS

The following stories represent firsthand accounts of the both the incidents that occurred on 10 October 2007 and 28 November 2007 from members of Task Force Bronco. The interviews occurred in September and October 2016. These are their words, and their stories have not been changed or adapted for publication.

Powerful Perspective for a Powerful Moment

Colonel Jon Althoff

> Rank and Position during Operation Iraqi Freedom 06-08: Major; Battalion Executive Officer (XO) and Brigade S1, 15th Personnel Services Battalion

In the beginning of the deployment I don't recall us having many mortar attacks until May, the big event happened after I was gone. I do remember on two occasions, on 7 December 2006 there was a big rocket attack. There were seven rounds and the last one hit Signal Hill. The other time was when me, the BC and the CSM were talking in the BC's office and one went right over heads and hit just outside the t-walls where we stayed. Another time one round hit the basketball court.

SFC Tobiason was a simple man with a heart of gold, I believe. That one, I will never understand what happened there. You talk about a lightning strike, bad luck situation for that guy. He was willing to do anything. He blended in well. I remember traveling around and telling the different detachments that they were all important to us and that we were all a part of the same organization. I remember telling them that we were going to work together to get through the deployment and that our goal was to get them home safely. I specifically remember the day I spoke with SFC Tobiason in the room and that still bugs me. Things happen for a reason and what I said there had no bearing on what happened later on, but when you are lying in bed and you can't sleep you think about things like that. I feel terrible about that and I always will.

When I took over in Minneapolis, I went down there and I found him right away. What I found was encouraging to me. It was obvious his gravesite had been visited regularly and I took comfort in that. I don't know who it was; if it was family, friends or the local VFW but I felt good about the fact that somebody was there.

First Lieutenant Angela Montgomery Capetillo

> Rank and Position during Operation Iraqi Freedom 06-08: First Lieutenant (1LT); Battalion S1 and Assistant Battalion Operations Officer (S3), 15th Personnel Services Battalion

Earlier that day SSG Hilliard dropped by the office and invited me to supper with them and I told them I would try. For whatever reason I decided not to go. I don't remember why, if it was a task I had to complete or what, but I just went back to my room. I got a knock on my door and it was SGT Williams telling me the rounds had hit by the DFAC and that LT Pinzon was there so I was the last S1. I was lost. Thank God SPC Castillo was trained in the reporting and getting whatever information we could. I don't know that we had learned the extent of SSG Hilliard's injuries yet, but we knew she was still with us. I can't remember exactly when we found out about Lillian (Clamens), but we did find out that LT Pinzon was alright for the most part. Someone came in at one point and told us to wrap it up and go to bed.

I honestly can't remember much about the next day, but it was very somber. The S1 shop had everything under control. I remember Pinzon taking a couple of days to himself. I think he and I were at odds a little bit and we got under each other's skin a lot but I do remember back in the office hugging him and telling him I was sorry.

I attended SSG Clamens' memorial service and I remember the somber tones. I couldn't believe we lost someone. I remember the shots ringing out. TAPS. I just kind of remember little bits of it. I remember the Commander's coming up and placing their coins. I remember I was stressed and I'm a crier, but I couldn't cry. I wasn't a smoker, but I went out and bummed a smoke and just sat and was quiet.

I visited SSG Hilliard at her home in San Antonio a couple of months after we got home. It was kind of surreal. She is so strong. You could tell she had struggled with it for a little bit but she knew she had to get better because she had two babies to take care of. She and her new husband reconnected and that really helped. She had support and she had someone that loved her.

I remember being ready to celebrate the BC's birthday and I don't remember when we heard, but I remember being told that another LT and I were supposed to go take care of SFC Tobiason's effects. Gathering his effects was hard. With SSG Clamens her belongings were already in bags to go home. But, with SFC Tobiason we had to go through and pack up all of his stuff. We prayed for his family. We drove his belongings over to Mortuary Affairs after inventorying it and packing it up.

Retired Major Rodney Davis

> Rank and Position during Operation Iraqi Freedom 06-08: Captain (CPT); Commander, Alpha Detachment, 15th Personnel Services Battalion

10 October was a beautiful day in Iraq, my 1SG and I was on the way from the gym heading back towards the Battalion area. It was around 1800 or 1900 hours, as we walked back towards the battalion we saw LT. Pinzon, SSG Hilliard and SSG Clamens out front, getting ready to walk to the mess hall.

I had made it back to my room and was about to it the showers, when all hell broke loose. Being accustomed to the sound of rockets and mortars, from OIF II, I immediately got between the concrete barriers by our hooches. Rockets flying overhead and hearing the sounds of impacts, I thought they were just shooting at "Signal Hill", a group of antennas on top of the hill by my office.

Suddenly, I hear that the mess hall was hit and we had casualties, I got on the phone and called my 1SG to get accountability because I knew we had Soldiers that were at the mess hall and the gym. Worried about our Soldiers and the Soldiers in the Battalion, I began to calm down and let the situation take place. I heard again later that LT Pinzon, SSG Hilliard and SSG Clamens were at the mess hall and may have been a part of the casualties. I immediately, ran to the TMC by the battalion and began to search for our people. I saw the Commander and she was looking for SSG Hilliard and I told her that I did not see her come in, not knowing that she was transported to the TMC on the other side of the base. I continued to look for her and one of the staff personnel came out and I told them who I was and whom I was looking for. They took me over to a body bag and opened it, I looked in the bag and noticed that it was a Soldier with braids, but it was not SSG Hilliard. I didn't realize or think to identify SSG Clamens because I had only met her once and it didn't register to me at that time. I remember the Commander and SGM asking and looking for SSG Hilliard once again and also looking for SSG Clamens. I asked what she was wearing and what she looked like. She told me and I took her over to the body bag, it was indeed SSG Clamens.

I could not allow myself to lose control and show any fear because my Soldiers were counting on me. I had been in this situation before as a Battery XO in 1/21FA during OIF II. I lost three Soldiers during that period and one of them actually died in my arms, he was only 18. Living through that period and remembering my training, I began to hunt for the others. My 1SG obtained accountability of our Detachment, we got in a vehicle and headed to the other TMC on the other side of the base. We got there and it was chaotic, people everywhere, looking for their Soldiers, friends. We waited there, seemed like forever. No sign of LT Pinzon, or SSG Hilliard. Soon after we saw LT Pinzon coming out, he suffered minor injuries, but was visible still shaken. I tried my best to console him knowing that there is no such thing when you have went through something so tragic. We also heard that SSG Hilliard was to be medevac'd to Landsthul hospital in Germany.

The BC asked me to fly to Kuwait and represent the Battalion. She wanted me to inform the rest of SSG Clamens' unit of what had happened. The BC told me "I need you to go and I know you can handle it". We got on the C130 and flew to Kuwait, as soon as I hit the ground I went and found the postal commander there and told him what happened and told him what Soldiers I needed to speak with. I got everyone into a conference room and had to break it to them at that point. That was probably one of the hardest things I've ever had to do. There was not a dry eye in the room. I was humbled. I was honored. And I was very, very emotional trying to deliver that message.

Retired Staff Sergeant Aaron Douglass

> Rank and Position during Operation Iraqi Freedom 06-08: Sergeant; Forward Area Support Team (FAST) NCOIC, Bravo Detachment, 15th Personnel Services Battalion

I knew both SSG Clamens and SSG Hilliard. SSG Hilliard because of being the S1 and SSG Clamens was doing the mail support and we would have breakfast just about every morning until I was picked up on the FAST (Forward Area Support Team) team. I wasn't around where the incident happened, but afterwards they told me LT Pinzon would be my new roommate because they wanted me to watch over him. It was a complete shock when I found out what happened.

When I talked to LT Pinzon you could tell it shook him because of what happened to two people he was standing there with. You see these people every day and then your world gets turned over and it's a completely gruesome sight to see. It's a shock to the system. I think he came out of it ok from my point of view but it's also one of those things that every now and again you'll see something and it will remind you of that person or that incident.

SGM picked me up from the flight line and it threw me off that he was there. He said we needed to have a talk and I was thinking my family or something like that. He told me everything that was going on and I just had that spaced out look. I felt like I was going to wake up from a dream. From there it was talking to other Senior NCOs that were around the FOB and let me know everything that had gone on. It was very emotional for everyone that was there.

Lieutenant Colonel Seth Graves

> Rank and Position during Operation Iraqi Freedom 06-08: Captain (CPT) Commander, Headquarters and Headquarters Detachment, 15th Personnel Services Battalion

I remember certain parts of that incident, I remember hearing the indirect fire, I remember taking the BC to the Camp Victory TMC after the attack, and I remember flying down to the CSH to see SSG Hilliard. SSG Hilliard I saw regularly prior to the incident. She was always positive and fun to be around. She was a good NCO, a good Soldier. I remember seeing LT Pinzon that night and he was very shaken up. But, that is all I really remember. The next day was sad. The Battalion had gone through a horrible night and all I really remember about that day was checking on Cristian Pinzon because he stayed in his room for a while. He was really shaken up by it and yet grateful to be alive all at the same time.

Sergeant First Class Steve Graves

> Rank and Position during Operation Iraqi Freedom 06-08: Staff Sergeant; Assistant Battalion Supply and Services (S4) Noncommissioned Officer (NCOIC), 15th Personnel Services Battalion

On 10 October 2007, everybody heard the incoming barrage of rockets, the explosions seemed endless. I was in the Battalion Headquarters (HQ) when the attack took place and we all ran to the bunkers outside and waited for the all clear. I can remember the EOC personnel doing their job to perfection as they had rehearsed so many times. When the smoke cleared (so to speak) I went inside to the EOC and checked their accountability roster against mine. When I saw the names of those unaccounted for, I asked SFC Hall where the rockets hit and he replied "The DFAC". I immediately felt sick to my stomach. I was asked to go eat chow with SSG Hilliard and SSG Clamens, but declined when my guys brought me back a take-out tray. There were several S1 folks in the DFAC at the time of the attack including their OIC 1LT Pinzon.

Retired Sergeant First Class David Hall

> Rank and Position during Operation Iraqi Freedom 06-08: Sergeant First Class (SFC); Emergency Operations Center, Noncommissioned Officer in Charge (NCOIC) and Detachment Sergeant, Charlie Detachment, 15th Personnel Services Battalion

I remember being in my hooch and I heard the first round hit and the alarm go off. I was still on shift with the Emergency Operations Center (EOC) and I was just about to go to supper so I had just stopped at my room for something. I remember that going off so my first thought is 'get to the office.' But, when I heard the second round and realized how close it was I realized I needed to get to a bunker. On my way I saw so many Soldiers running and so many that were scared and actually lost right there by the bunker and about to run back the other direction because they were so scared. When I found a bunker I stuck my head out and shouted at some Soldiers, telling them to get to the bunker. The rounds were like footsteps. The first hit over by the DFAC and they slowly kept moving forward until I was sure one was going to land right on us. I remember Soldiers asking me a bunch of questions I didn't have the answer to so I tried to stay as calm and collected as possible. I tried to not act scared so that maybe they wouldn't be so scared. I remember one Soldier in particular that was literally so scared they had balled up on the ground in the fetal position, crying and saying they were going to die. I got them up and got them up against the wall of the bunker.

I remember the last round hitting and realizing nothing else was coming. So that's when I told the other Soldiers to stay there until the all clear and I headed back to the office to figure out what was going on. I don't know how Sergeant Major (SGM) found out our Soldiers had been hit, but I remember him coming back, out of breath and asking where the BC was at. He told me to get accountability and then to hold down the fort and he said he would be back when he could be.

I've been to a few memorials; what I remember most was how well put together the service for SSG Clamens was. The way the video and the pictures and the music and just the way the story was told, it wasn't a celebration by any means but we were looking at the brighter side of her life. I remember the lump in my throat. At the same time, I remember hearing people laugh too because of the pictures and the video.

John was in the EOC with me when he first got over there so I saw him and talked with him every day. He brought laughter to our team. He took the edge off. I can't do nothing but smile every time I think about it. He's one of those people I wish I could be like. I know in the beginning there were some questions of his abilities, but he showed he could do it. He got a handle on things and everyday he had something good to say.

For his memorial, I remember SGM chewing my butt because I had been put in charge of the firing squad for John's memorial and there was an NCO who had done several of these that they sent my way to help me with anything I needed. But, I wanted it to be perfect and I had participated in others at other places throughout my career. The NCO tried to tell me how to do a few things and he asked if there was anything else I needed and I said "no, I think we got it" and somehow it got back to SGM that I had run the guy off so it put kind of a stain on the whole thing for me. It's not the first time I've seen death and I try to be very stoic during the whole process, but I remember I had a few tears roll that day. I enjoyed the video presentation with the pictures and the music. The only thing I can say is that I miss him.

Retired Sergeant First Class Ireshekia (Hilliard) Henry

> Rank and Position during Operation Iraqi Freedom 06-08: Staff Sergeant (SSG); Battalion S1, Noncommissioned Officer in Charge, 15th Personnel Services Battalion

I don't recall the attack much. Some things I remember and some things I don't. I was told to go to the Sustainment Brigade to pick up awards, but when I was supposed to fly out the flight was changed. I was supposed to leave earlier in the day and for some reason that flight wasn't going out so it got pushed back to later that evening. I was happy because I would be able to see at least some of the Gospel Revival and I knew SSG Clamens was leaving so I was like "let's go have dinner one last time". She wasn't fond of leaving her room but I told her I wasn't going to Burger King that we would go to the dining facility. After that it is like my memory was erased.

My very first memory (after that) was waking up and asking if my Soldiers were ok. They told me they were fine and at that point I didn't even realize I had an injury. I remember waking up a lot and not remembering what I had said or what anyone had said to me. It wasn't until after several times of waking up and going back to sleep that I remember waking up and looking down and seeing only one foot sticking up. I realized it was just my foot, or a little bit above the ankle, and I cried a little bit and that was it because I realized it could have been a lot worse. There are some things I have remembered and I don't know how I remember it, but I knew my BC so well and I remember her telling them to do whatever they could to save as much of my leg as possible.

I have a picture of SGT McKinley, SGT Graves, SGT Castillo, and someone else in the picture with me in the hospital room. Someone gave me a camel and I still have it in my room. I also still have the bag of my stuff from that day. I came across it once and put it back; I know it still has my favorite watch and my nametag and my clothes with our blood on it. It has been sealed and I have not gone through it yet.

The day I got my Purple Heart, there was a big wig standing over me and asked me if I knew who he was. I looked up and read his nametag and said "yes!" because it was General Petraeus. I remember him pinning the Purple Heart to my shirt.

I went from Iraq to Germany, Germany to DC, and then to BAMC (Brooke Army Medical Center in San Antonio). In DC I was at Walter Reed Medical Center but I don't really know why I was sent there. I think it was just generally to treat me for my wounds. The shoulder wound I didn't even realize I had but they had to do debridement on it and on my leg so I know I was there for at least one day. That was when I was introduced to the Wounded Warrior Project. When they were cleaning my wound I had no idea I had sustained such an injury. I turned to look at it and I gasped because I didn't realize it was so bad. I stopped moving around but they told me to keep moving because that would help it.

I remember using CPT Graves' phone in Iraq to call my (twin) sister. I couldn't give her details, but we have this closeness and she really does feel my pain. I remember getting choked up and all I could say was "sister", when I said that she knew something was wrong. I handed the phone back to CPT Graves and he wasn't sure what to do because he couldn't reveal much information. I remember him telling her "all I can tell you is that she is ok and she will be on her way home soon." My sister, out of the blue, she asked him "is it her leg?" because she had started having pains. We are that close. All he could do was say "yes, she's fine" trying to answer her but not answer her. My sister pretty much knew. When they were wheeling me into BAMC the first person I saw was her. She stayed with me; she literally stayed in the room with me.

I did not know about SSG Clamens until 1SG Smith called me after I had been at BAMC for several days and he told me the service was nice and how the battalion did a really good job and I asked him what service was it. He realized I didn't know but he was past the point of no return. That's when I found out SSG Clamens was killed.

When the battalion re-deployed they said if at all possible I could be there when they got home. That was a big deal for me. I definitely wanted to be there. This was my family. I didn't have to think about if I wanted to be there or not. I definitely did. It was an honor to be back with my unit and it allowed everyone else to see how I was doing because there were a lot of Soldiers and friends who didn't know how to approach me and would avoid calling me because they didn't know what to say. I remember trying to stand briefly at the end of the homecoming ceremony.

People assume when they see me and then they know I was injured in the military, one of their first assumptions is that I was doing something that required me to be outside the wire. They ask if I was Military Police (MP) or something and I tell them no, that I was personnel. So then I get this look. People have this perception that only those in the combat positions are in danger. I tell them that nowhere is safe in a combat zone, that's why they call it a combat zone. That is a shock to a lot of people. Everyone over there is in danger and you never know where it's coming from.

When I was injured, one of the bits that I do recall, I remember tasting dirt and I remember knowing that I was not alone. At the time I didn't know why I felt that way, but then I was later told from some of the medical personnel who had helped me that I was able to visit, they told me how they had to cut everything off. They told me they were trying to assess my injuries and they had to remove my clothing because SSG Clamens and I were so close they couldn't tell what my injuries were because I had a lot of her blood on me. One of the Soldiers in particular from the medical unit actually came to visit me at BAMC while she was home on leave. She was still in shock somewhat and she walked into my room and I didn't know who she was and she started to cry. She told me she knew I wouldn't remember her but that she was one of the medics who helped me. She said she had been in the DFAC and had run out when the attack happened. She told me she gave me morphine because she didn't want me to feel pain and I didn't throughout the entire ordeal. She was crying and told me that she didn't think I would make it so she was so happy to see me. I later talked to SFC Malone and she told me that she navigated her way through several Soldiers that day and she didn't know why until she found me, not even knowing it was me, and rolled me over and was able to recognize me.

There was another black Soldier that was injured in almost exactly the same way. She lost the same leg in almost the exact same spot and I remember her telling me when we were in the hospital that she felt so alone and I remember telling her that I couldn't share any of that with her because I didn't go through that. I was comforted. I never felt alone.

Lieutenant Colonel Dale Jones

> Rank and Position during Operation Iraqi Freedom 06-08: Captain (CPT); Commander, 175th Adjutant General Company (Postal) (Direct Support)

Not an October 10th goes by that I don't think of Iraq. On 10 October 2006 we had the rocket attack on FOB Falcon that caused the Ammunition Holding Area (AHA) to explode and 1LT Falcon received a Purple Heart (PH) from that. Then the 2007 events which I was talking to Shawna on the phone over at the office, before heading to the DFAC. I heard the rounds coming in and dove under the table but left the receiver on the desk. The sixth or seventh round hit close to the office and Shawna could hear it over the phone, she had a much harder time with me being there after that. Calling her and getting through kept me away from the DFAC and being in the middle of the much worse impacts.

Retired Master Sergeant Patrice Moses

> Rank and Position during Operation Iraqi Freedom 06-08: Master Sergeant (MSG); Detachment Sergeant, Bravo Detachment, 15th Personnel Services Battalion

I interacted with SSG Clamens daily. As a 1SG I would make my rounds and check on my Soldiers, and then she was assigned to the S1 for a while so I was interacting with her even more. And then there was problem with their leadership so she was put under us and she was a squad leader so for accountability we interacted often. After duty hours we would talk about personal things like her girls. She was very positive, always upbeat and a go-to type Soldier where if I needed something from the postal side, I knew she was a person I could go to and get it done. She had a take-charge type attitude.

The night of 10 October 2007 we got a phone call. I can't remember how exactly I found out about her, but I remember having a conversation for whatever reason with SGM Edwards and she told me to give it a few minutes and then to call and check on my husband. When I finally got a hold of my husband, he was the one that told me about SSG Clamens. And I said no it can't be because we haven't heard anything from the battalion. I didn't want to say anything until we got confirmation from the battalion. Shortly afterwards we received the information and we had to figure out how we were going to tell the Soldiers and we were waiting on the Chaplain. It wasn't until early the next morning when we got everyone together.

Retired Sergeant Craig Morrison

> Rank and Position during Operation Iraqi Freedom 06-08: Specialist (SPC); Emergency Operations Center, 15th Personnel Services Battalion

On 10 October 2007, SGT James Zimmerman, SGT Wayne Keller and myself had just left the DFAC with our dinners (take-out). We literally just got into the Battalion runner truck when the sirens started going off to alert everyone of impending incoming. There had been a lot of false alarms recently, so we all kind of hesitated. Then the first of I think were seventeen rounds hit our Area of Operations that night. We all jumped out of the truck and started heading towards the nearest bunker. Then a round hit what seemed like a very short distance away from me. I can remember seeing the sparks fly just like you'd see in the movies. I also remember seeing the silhouette of people running between me and the explosion. I just stood there in awe of what I just saw. Then I felt the pull on my arm from SGT Zimmerman and we ran to the bunker where SGT Keller and a civilian worker form the DFAC were already seeking shelter. I can remember on the way to the bunker hearing the hiss of the rockets flying overhead and hearing the explosion hit near and afar. My back had already started hurting me at this time, so I remember that my running probably looked quite comical. When I look back, I sometimes wonder which of the explosion I heard that night was the one that affected us the most. I never got to know SSG Clamens very well. We had only talked a few times since she was not stationed at Camp Liberty with us. I knew she had a family back home and all I could think about was how I felt so sorry for them.

When the 487th joined us, SFC Tobiason was put in with us and he worked with us in EOC for a number of months. What I remember from him is him helping me study for the Board. Everyday that's what we did; we studied for the board. I contribute a lot of my passing the board and getting E5 (pay grade for rank of Sergeant) on him because he helped me study every day. We ate together at least two meals a day. He was a double cheeseburger and French fries guy, that's what he'd eat almost every day; two double cheeseburgers and French fries He was very funny, a super nice guy. He always had a smile on his face; I wish I would have known him in civilian life.

That day I got to the EOC and both doors were shut, usually at least one was open. So, I opened one slightly and peeked in and Staff Sergeant Martinez was on the phone and I heard him say something about a KIA and then when he said "SFC Tobiason" I can remember sitting in a chair, but I don't know how I got to the chair. It hit me hard. He had been gone for a few months back out to BIAP but I had literally just talked to him on the phone either earlier that day or the night before when he called in for accountability. I just couldn't believe it. They said a stray round came in from off base somehow and hit him, I guess, and he was standing in line to meet the Washington Redskin's cheerleaders. He got struck and he went down and that was it. He was killed instantly.

The two things about that night that really hit me hard was that I volunteered to clean his weapon because his weapon was bloody, and then that night when I was walking back to my room I saw SFC Hall sitting on his porch drinking a non-alcoholic beer and sitting there talking to him for a few minutes was pretty rough. As for the ceremony, I can remember it from beginning to end. The hardest part was seeing SFC Hall cry and seeing him get down on his knee and grab SFC Tobiason's dog tags, that was rough.

I just know that day for me, by far, was the worst day of the deployment.

Chief Warrant Officer Two Nicole (Behrens) Patterson

> Rank and Position during Operation Iraqi Freedom 06-08: Private (PV2); Battalion Awards Clerk, 15th Personnel Services Battalion

I remember it being very often, at least three or four times a week to the point where you just got used to it. Which is a sad thing. "Oh it's just a mortar". It was frequent enough to where we became numb to it. That day in particular, the 834th was coming through BIAP on their way home and they were going to do their awards ceremony in Camp Liberty. I remember work as usual but we had SSG Clamens there in our office for most of the day. It seemed like a joyful day.

I remember I got off work at the regular time and I was going to go to dinner. I went to dinner with my roommate and rather than working out first we went to dinner first. We decided to skip the ice cream and leave. When we were leaving we were passing the gym when we ran into LT Pinzon, SSG Hilliard and SSG Clamens. We said hi and I remember laughing about something. By the time I made it back to my room we heard a mortar. We had a bunker right next to our trailer so we immediately went into the bunker. 1SG Hall told us it hit the DFAC and we're trying to get accountability. I remember waiting and wondering and it was awful because I knew people who were on their way there. It was dinnertime so anybody could be there right now. I was told I needed to report back because we had casualties. I don't remember who told us who was involved, but someone told us LT Pinzon was being treated for wounds, SSG Hilliard was alive but she had been hit pretty hard, and then I read in a report later that SSG Clamens had passed.

I remember being told that SSG Hilliard was being transported to the Green Zone and she wanted to know how her Soldiers were, so the next thing I knew I was going with the BC and the CSM and several others to the Green Zone. There was a waiting period so I remember sleeping while we waited for her to get out of surgery. It was awful. When we walked in, she was covered in blankets so I didn't see that her foot was gone, but her face was puffy and she was very groggy. She wasn't SSG Hilliard. I mean it was, but it was so difficult to see my boss look like that. I so badly wanted to close my eyes and have it be different, have it not be happening. But, it was awesome because she had a smile on her face and she was the one trying to cheer us up, making jokes and trying to make us feel better. I felt guilty for being sad because she wasn't. I think she felt lucky. God was with her holding her hand and she made it out alive. I remember escorting her out of the building so we were there with her until she left for Germany. I'm glad the BC helped make it happen so that I could go visit her. As a Private I wouldn't have been able to make that happen on my own. I feel so grateful for being able to be there. It was good to know that I got to see her off.

I remember talking to her on the phone, but I didn't see her again until August 2008. I had moved units and I asked SSG Hilliard to be at my promotion ceremony when I was promoted to SGT. I wanted her to be there to pin me. She was my role model from day one. She was the reason why I had most of those points because she wouldn't let me have a full hour and a half for lunch if I didn't get my work done. I had a lot to thank her for. If I wasn't enrolled in two college courses at a time she would call me out on it. She pushed me very hard and I have to thank her for that.

After being treated for his wounds, LT Pinzon came back to work. He had a different spirit after that. He was changed. I can only imagine why and how and I don't know what he saw or how much he remembers, but he was different. He was very, very, very much more compassionate and personable and caring. I think he was literally thankful for every moment that he got to have after that. He didn't want to take any moment for granted.

SSG Clamens was actually responsible for my ex-husband, who was a troublemaker, on a daily basis. He was very controlling of me and needed my attention all the time. I would have to call him in the morning and in the evening, every single day, so I would talk to SSG Clamens quite a bit. She gave him and me some advice sometimes and she would say that she would pray for us.

I did the video for SSG Clamens' memorial of pictures I received from everybody and so we had a beautiful video with music to remember her by. I also did the memorial program. I remember the roll call more than anything. As respectful and great that the military makes memorial services, I wouldn't wish for anyone to have to go to those. During roll call when they call out everyone's name and receive a response until they reach the fallen, it is just so hard. Everyone goes up and salutes and pays their respects. It's an experience that shouldn't be had by many.

The only thing I remember is being informed of SFC Tobiason's death. There were some NFL cheerleaders visiting and a lot of the guys wanted to go meet them. I was told he and a couple of friends were outside smoking and there was a firefight on the other side of the wire and a stray round had come over the fence. The people that were with him didn't even realize it had hit him until he dropped. I know SFC Tobiason extended his tour in Iraq and that happened during his extended period so the rest of his unit was gone by then.

At his memorial service, when they did the roll call and they called out every Soldier and they say 'present' until they get to the fallen Soldier every hair on your body stands up. Even with SFC Tobiason, I wasn't all that close to him, but during his roll call his voice wasn't in the room with us. It's heartbreaking.

It was another sad day for the 15th PSB.

Chief Warrant Officer Two Yvonne (Adams) Ralph

> Rank and Position during Operation Iraqi Freedom 06-08:
> Private (PV2); Personnel Actions Clerk
> Charlie Detachment, 15th Personnel Services Battalion

The area of focus would be 10 October 2007. The morning of 10 October 2007 I was extremely excited because the FOB was hosting a gospel extravaganza that evening at the basketball gym on Camp Liberty (across the DFAC). I had just arrived to the extravaganza that evening when I heard the first mortar round. It had to been sometime after 1800 I think. Right after the first mortar round went off several more followed. I immediately yelled out "OH SHIT" and ran to the bunker outside the basketball gym. I remember we were all trying to make room in this small bunker and praying out loud. I do not remember who was with me I just know it was about 20 people near the bunker I was trying to get in and over 100 attending the extravaganza.

Once all clear was given, I walked back to the unit on PAD 11 to ensure my chain of command was aware I was ok. On my way there I remember seeing several individuals with blood on them and many others running. I was so scared.

Major Richard (Rick) Strong

> Rank and Position during Operation Iraqi Freedom 06-08: Captain (CPT); Commander, Charlie Detachment, 15th Personnel Services Battalion

On 10 October 2007, we had the daily meeting over the computer and we had an end of tour awards. The rest of the day was a typical, normal day and then we were at the ceremony that night. It was about sunset. Right after the ceremony a bunch of folks were going to walk over the DFAC. I didn't go. I went back to my trailer and the alarm sounded and rockets came in. We got out and went to the bunker there. I went to the one that's right behind the Battalion Headquarters. There was a bunch of people in there. I wanted to see what was going on so I left the bunker and went to Headquarters with SPC Castillo and we found out there were casualties. Because my team had CLTs that did casualty reports all the time I was familiar with them so I was in the S1 trying to help them. I was trying to calm them down a bit and walk them through the casualty reporting process. I don't think anyone was deceased at the time but I knew there were some with more serious conditions than others. I remember after the incident hearing about exactly what happened and hearing about Hilliard, but at the time it was happening I didn't know any of that stuff.

Colonel Timothy Zimmerman

> Rank and Position during Operation Iraqi Freedom 06-08: Major (MAJ); Executive Officer, 15th Personnel Services Battalion

We were going about the day like any other until the alarms went off. We had all talked about going to chow that day with SSG Clamens. The S1 section, Pinzon stuck his head in and invited me. There was something I was doing and I said "no, not right now but I might catch up with you all." By the time I started that direction everything happened. To this day I have no idea what I was doing but there was something I had to get done, otherwise 9 times out of 10 I would have gone to chow with Pinzon or Deakyne. Those sessions they actually relaxed a little bit and you could actually talk to them. It was a great opportunity to do mentor sessions and things like that. This one particular time it was like a sendoff dinner. But, there was something I had to get done and I didn't go with them.

I remember the event happened and I was immediately back at Battalion Headquarters working with a Sergeant trying to get accountability. At first it was kind of sketchy, we heard there were multiple casualties. We narrowed it down to SSG Clamens, SSG Hilliard and LT Pinzon but there were others at the DFAC. We tracked down the BC and she and CSM went to the TMC and CSM would call back in and give us information. Once we heard SSG Clamens had passed we transitioned immediately into the memorial planning.

While we were still in country LT Pinzon had a really hard time talking about it. I think he had a lot of survivor's guilt. What happened to SSG Hilliard affected him a lot.

1775
THE JOURNAL OF THE ADJUTANT GENERAL'S CORPS REGIMENTAL ASSOCIATION
SUMMER 2007 EDITION

1775 Presents a Special Feature on the 15th PSB in Iraq

The following articles were written by members of Task Force Bronco during our deployment to Iraq in 2006 & 2007. These represent the actual words of my Soldiers and I am honored to share them with you here.

These articles have been reprinted with the full permission of 1775.

Why I Serve

Written by: 1LT Angela M. Capetillo, Assistant S3, 15th PSB

In spring of 2002, I was in the Delayed Entry Program waiting for the school year to end, so I could attend Basic Training. However, I wasn't a typical high school student enlisting in the Army; I was a high school teacher enlisting in the Army.

At the ripe old age of 26, I had already graduated college and had been teaching high school for two years when I decided that teaching really wasn't what I was meant to do. I had been doing some research and decided to be a flight attendant. I went to an open interview and didn't get called back. I noticed that two of the applicants had been in the military. "Hmm....," I thought.

I signed up to be a 71L Administrative Specialist. It was what sounded closest to my business degree. I had no idea that it belonged to the AG Corps or even what the AG Corps was until later on at AIT.

I attended Basic Training and AIT at Fort Jackson. I learned to type memos, use the MARKS filing system, and how to use Form Flow. Then they asked me what I thought to be a strange questions.... "Do you want to go to F5 School?" Huh? Someone explained that is was postal school. At that point, September 11 had just occurred and I was so homesick that I turned it down. I just wanted to get home to see my family before going to Germany for my first assignment.

What do you know, I ended up in 510th Postal Company. Couldn't they have told me that before I turned down postal school? Luckily, I was on the operations side of the house versus the finance side. I quickly found myself in the Registered Mail Room being trained by a PFC that I was soon to become the "SIC: Specialist-In-Charge" over. I caught on quickly and became quite proficient. Like any other Soldier, I had days that I thought my platoon leader was an idiot. Boy, I wish I knew then what I know now….. More to follow.

During my assignment to the 510th Postal Company, I found myself deployed to Kosovo where I spent my six months on Camp Able Sentry across the border in Macedonia. I never knew until then how important the postal mission was. There were only three of us working the APO and everyone on the camp knew who we were. I was once referred to as the "Postal Chick." Needless to say we were well taken care of.

It was during this deployment that I made the decision to go "Green-to-Gold" with some urging from LTC Marlow, one of the kind souls at Camp Able Sentry who took interest in the postal Soldiers. I had felt for some time that my education was going to waste, and with a little prodding, I decided that I would work on my MBA while making the transition to the Officer Corps. Besides, maybe I could make a difference and do it better than the one lieutenant that I thought had lost his mind most of the time with some of the ideas he came up with.

In the fall of 2003, I became a member of the Texas State University-San Marcos ROTC program in addition to being admitted to the graduate program. I found my rhythm going to PT and /or ROTC classes in the mornings and did my homework while I hung out waiting for the evening graduate classes. I spent about 45 minutes on the Interstate commuting to and from home and for a couple of semesters, I had a class at a satellite campus another 45 minutes from the university in the opposite direction from home, making the drive back home at night an hour and a half.

Ultimately, I assessed into the AG Corps which was my first choice, and was commissioned in August of 2005, one class shy of my MBA. I attended OBC a month later and found myself at Fort Hood in the 1st Cavalry Division by January 2006. I was assigned to the 15th Personnel Services Battalion and became their S1.

It didn't take long for me to figure out that being an S1 in a battalion full of AG officers was a blessing as well as a curse. All the knowledge I could ever want was at my fingertips, all I had to do was just ask. In the same turn, everyone had done it this way or that way which was better than my way. Luckily, I had a great NCOIC, SSG Hilliard, and a great group of Soldiers to teach me the ins and outs of how they ran the S1 shop. I've learned plenty, and I thank them. I also have had two great XO's guide me. MAJ Holbrook who made me dig in and take that last graduate class, and now MAJ Althoff; I thank them for their patience and words of wisdom.

The 15th PSB has now been deployed to Iraq for a few months, and I've moved on to being in the Battalion Assistant S3. I'm in a combat environment learning to track air movement requests, and learning a broader view of the Personnel Services Support mission, and a new set of skills to put in my kit bag to make me a better officer.

As I've developed as a leader, I've learned that, although I ultimately don't like to run, it can be a great release for stress and frustrations. I've now gone as far as joining a running club started on our camp to train up for a half marathon. I'm no PT stud, but I've come to enjoy the fact that I can now run over 10 miles. Mostly, I enjoy when it's over. If any of my old enlisted buddies were told that I was running this much, they wouldn't believe it and they would tell you that you were lying. I love the fact that I have Soldiers who ask, "So, LT, how far did you run this weekend?" I tell them and get a "Way to go, Ma'am!"

Now, I'm coming up on the downhill slope of the deployment and have to look to the future of my career as an officer. The 15th Personnel Services Battalion will be inactivated. Where do I want to go next? Will I get that choice or will it be chosen for me? I've thought of starting a family. Will that affect the career choices I make? I have faith that whatever comes my way will make me a better officer, leader, and person in the end. And, what if down the line I decide that I want to join the civilian world again? I think I will be fully prepared for what that world could throw at me, thanks to this experience.

So why did I tell this story? I just wanted to share my story....we all have one. I learned that what I thought of officers as an enlisted Soldiers was all wrong. I figured out that maybe the lieutenant I was trying to outdo, was just doing the best he could with the knowledge he had at the time. I figured out that being an officer is just like anything else in life; it is a learning process. There is no way I can have all the answers all of the time. I realize that when I get a funny look from a Soldier, it may be because the idea I came up with is the stupidest thing they have ever heard, just the way I thought of my LT when I was enlisted. Then, I have to re-think the issue and look at it as if I were a Specialist again.

Finally, I've learned that by taking this journey, not just from enlisted to officer, but from civilian to Soldier, I've accomplished something that too many of my 'Generation X' won't be able to say. I've served my country, I've made my family proud, and I've had the honor of having a member of the 'Greatest Generation' tell ME "Thank you for your service," How cool is that?

Why I Serve

Written by: CW4 Ronald Mavity, Detachment 4, 847th Personnel Services Battalion

They joke today about me being the "old man," not only because of my age, but because after 34 years in the military it is hard not to be called that. I would like to take the long route in telling you why I serve, because some decisions are not made overnight nor even with clear reason. It all starts back in the summer of 1972 when I was at college, and, with every other male of my age, I was watching as dates were selected for the military draft. I know most of you do not understand how this works, so let me explain. A large bowl is filled with individually wrapped dates to cover every day of the year (i.e., January 1, 2, etc.). These will correspond with the birthdates of every male of age for that year. As you can imagine, all of us watching that day had a vested interest in what dates were pulled. You see, whatever dates were pulled first were sure to go to Vietnam. What actually would happen was after every date was pulled from that bowl, the order of the dates was written down. Then, congress decided how many numbers were needed (i.e., first 50 or 100), and everyone would know if they were going into the military. Of course, as organized as the government is, they don't tell you immediately how many numbers they are going to take. So, if you were #95, like me, you were wondering even more. That day brings back mixed feelings, someone in the crowd said, "Oh that's my date." Reality hit quickly!

By December, I had had enough. I had been at college for three semesters and was plummeting down the "away from home, do nothing attitude" and with poor grades. I was definitely not getting anywhere, but I had a real problem. I was not going to Vietnam if I had anything to do with it. Like today, the media gave body counts every day and told us how stupid it was to be there. At 18 years of age, anything that felt good was good and anything that would eventually hurt me was out of bounds, so going to Vietnam was out of bounds. I started looking where I could go for a very short period and get this whole military thing over with. Now remember, if you were drafted, you were enlisted for one year, but you definitely had to go to war, so I went to the Navy recruiter. My father was in the Navy after Pearl Harbor, so it would be good for me too. Well, this guy tells me I would have to stay four years. Are you crazy? I went to the National Guard (I didn't know the Army Reserve existed). The guy says he could probably get me in even though they were over filled, but of course there was a fee. A fee? Yes, if I would give him this amount of money, he would make sure I would get in his unit. Yes ladies and gentlemen, that was happening all the time. Next, I went to see the Army recruiter. He says I'll sign you up for six years, but you only have to do three; we have this new thing called the "All-Volunteer Army." And by the way, you can pick your job and your first assignment. Well I had always been fascinated with history, and a specific college professor had fascinated me with European history. So I said fine, I am a mechanic and I want to go to Germany. No problem (for him anyway). When I got to the MEPS, all of a sudden there were no training seats for mechanics, but if you are a truck driver, you can work on your own truck. I said no I want to be a mechanic. Sorry can't help you. Well, being young and naïve, I thought this was my last chance. There was no college waiver anymore to get you out of the draft. You couldn't even get out of it if you were a sole parent. The recruiters and MEPS personnel basically had all the power, so I said yes, and on 20 December 1972, I joined the Army.

I went to basic at Fort Leonard Wood, Missouri and then went to AIT at Fort Ord, California. This Army was 95% male, cursing, scratching and physical. No stress cards. If the drill sergeant hit you, he must have had a reason, and yes it happened all the time. When I got to AIT, I got very sick. I ended up in the hospital, and then I started having epileptic seizures. The Army wanted to get rid of me, to say the least, and I, and a congressman wanted me to stay. I won, I think. They reclassified me as a 71L, Administrative Specialist. I then went to Germany and to Headquarters 7th Army and European Command, War Plans Branch of DCSOPS. Wow, I was a PV nothing, and the lowest ranking officer was a Major. It was a completely different Army then. Men ruled, and they all had combat time. They smoked at work, and none of them did PT. And you better believe they were one grade below God, or else you were doomed to spend eternity as a private. I learned a lot and after my there year tour was up. I returned to my wife and college.

Now remember at this time in my life, I am happy to be out of the Army. I am definitely more disciplined, but I don't want any more of that Army stuff. Even though I had the GI Bill, VA compensation and everything else, I heard about a thing called the U.S. Army Reserves. Honestly, I didn't know they existed. I needed new tires for my car. Things like tires were extra. School, food, clothing, and then extras; if there was money. I found out that as an E-5 in the Army Reserve, for one weekend I could get $135. That could buy some tires, so I joined and became their new Supply Sergeant. Funny thing is that I didn't see any of that money for six months.

Life goes on, and after sixteen years with the Army Reserve, many summer camps, children, houses, and numerous car tires later, I heard about a new program for selecting Warrant Officers. Now my interest in Warrant Officers goes back to my first day in Basic Training. Our group was gleaned of a certain bunch and we were taken to be tested. We were later told that it was a test to determine if we would be selected for flight school and to become a Warrant Officer. I didn't even know what a Warrant Officer was, but over the years I learned and became more and more interested. After eighteen years, though, I was old by Army standards; but I did not only put in one packet, I put in two. I was selected for AG Warrant and set to Fort McCoy Wisconsin in 1988. That experience was a harrowing one. Not only did they push you physically, but they required you to change – to change the way you act because you were no longer an Enlisted

Soldier, to change the way you act because again, you are now a Warrant Officer. Harrowing but exhilarating. After Ft. McCoy, I went the next year to Fort Benjamin Harrison, Indiana and was pinned a W01. I have had many wonderful and exasperating assignments as a Warrant, not only as a drilling reservist, but also as AGR.

Being a person who has come from being willing to protest the war in Vietnam to one who serves in Operation Iraqi Freedom is easy to explain. As I said earlier, as a young man I was concerned about what felt good and basically what pleased me. I took someone else's (the media's) opinion about what it was like, and agreed that the military was not for me. Then, I experienced the real thing. The camaraderie, the learning environment, the ability to advance at your own pace, and basically that ownership of MY ARMY, made me not only belong but made me desire to stay and serve. Now, I know the difference between what someone else says the Army is and what I know it is. That is why I serve, because this is My Army.

Why I Serve

Written by: 1LT Dexavior Smith

I entered the U.S. Army in June 1992, immediately upon finishing high school. I was 19 years old and I enlisted as a means of gaining my independence as a man. What was very motivating also, was the fact that they offered thirty thousand dollars to attend college. I knew that I wanted to attend college, but being raised in a single-parent home, my mother didn't have the means to support my goal of attending college. I figured doing 4 years in the U.S. Army, learning a technical skill, traveling, gaining my independence, and earning thirty thousand dollars towards college wasn't a bad way to start my life after high school.

I went to Basic Training at Ft. Leonard Wood, Missouri, did Advanced Individual Training at Aberdeen Proving Grounds, MD, and was assigned to Fort Bragg, NC for permanent duty. A year after arriving at Fort Bragg, my daughter Kambrelle Denee' was born, so many things had occurred in that small block of time. I had developed, grown, learned from life's experiences. Before I knew it, four years had come and gone. I had met many people, some bad, and some good. I still remain friends with some of them to this day.

Upon finishing my four-year term of service, I took the opportunity of attending college. I was still listed within the data banks of the (IRR) Inactive Ready Reserve, which meant in the case of the Army's needs, I could possibly be called back to serve. There was no call, and after a year had passed, I felt as though something was missing. After consideration, I realized that the things I missed were putting on the uniform, the camaraderie amongst Soldiers, traveling to different places, meeting new people, and the periodic break in monotonous routine of everyday living. My next step was enlisting in the Army Reserve and, although it would be only be once a month and two weeks during the summer, it would give me the opportunity once again to experience what I was missing while still finishing college. Soon, my junior year of school was coming and I decided to apply for an ROTC Scholarship. I won the scholarship and then I was on the way to becoming an officer in the United States Army. Wow! It's amazing how things can end up in life sometimes. Three years after winning the scholarship and graduating from college, I became a Second Lieutenant in the United States Army. So, here I am, a teacher and an officer in the U.S. Army, which makes for quite an impressive resume.

The look of respect you immediately gain from others when they hear you are in the United States Army, the sense of pride you have walking down the street in uniform, the feeling of knowing that you belong to one of the strongest military forces in the world – it is because of you that the world is safe. We as an Army fight afar, as opposed to fighting on our back door step to keep our country safe from harm. We have been highly trained, prepared mentally and physically, and given a mission to complete. I voluntarily signed an agreement to do all I can to protect, serve, and defend the Constitution of the United States. I understand the commitment that was made; I must adhere to it regardless of whether or not I agree with the circumstances, and above all, it is my duty.

It is my honor and duty to serve with dignity as I know, "Injustice anywhere is a threat to justice everywhere," said Martin Luther King, Jr. As a powerful nation, we are charged with the responsibility of ensuring what is right amid a world of indifference.

Why I Serve

Written by PFC Christina M. Malpass, HHD 15th PSB

Why do I serve? It started when I was in the first grade when my Dad enlisted back into the Army. I've dreamt of being in the Army ever since. I moved to Fort Stewart when I was sixteen and my Dad's rotation of deployments started. I knew he was doing good things for our country. I looked up to my Dad, admired his strength to go, and his mentality to keep it together for a year. This made me proud to call him my Dad. I wanted to be there next to him, showing that I could do that, and do great things for my country also. By the end of my senior year I realized I wanted to serve my country and to see what I was really made of. I knew I could go to college and cheer but I wanted to see if I could make in the Army, if I could be as strong as my Dad. Why go and do something that just about everyone does by going to college, instead of doing what the elite few do and join the Army in a time of war, in a time when your country needs you the most?

I enlisted into the United States Army September 20, 2005, to be a 27D, Paralegal Specialist. My friends tried talking me out of it; they asked me why I would want to do such a thing. They told me I would deploy and asked me if I was crazy. My answer to them was, I enlisted because there is no better job that you can have that has all of the benefits that the Army offers. They give you a place to live, including utilities, they feed you, do PT every day to keep you in shape, and also give you opportunities to travel. I told them that it was like any other job, you have consequences to everything that you do. I told them the Army was perfect for me. When asked if I was scared of deploying or dying, I answered, naturally everyone is scared of death, but my outlook on it is, if it is my time to be called then I will go regardless of where I am or what I am doing. I want to deploy to help those who want help, everyone deserves a fighting chance. I want to help the children of Iraq who know only the ways they have been taught.

I wanted to come over here to see history for myself rather than watching it through some propaganda on TV. I wanted to make a difference and make my footprint in history. I wanted to see what I was made of. I serve my country to make a difference, be it big or small; I make a difference. I am not like the people at home talking about the war and doing nothing; I am here doing something about it. I want to see a change for the Iraqis and to help with that change. I know nothing changes overnight, but that all great changes come with time, and I want to be here to help make those changes. I serve because it is my calling to do so.

Why I Serve

Written by: 1LT Robert W. McAdam, XO, 834th AG CO (postal) (DS) during OIF 06-08

After fourteen years of service, this question is still easy to answer. Honor, duty, loyalty, and just wearing the uniform! It sounds kind of hokey, but through the years as my career progressed, on occasions when things seemed unpleasant or I wanted out, those three words and that phrase "Wearing the Uniform" still made me proud to serve.

At the age of 21, I woke up one morning in February 1993 and decided to take a drive to the nearest recruiting office to pronounce my loyalty to the United States Army. Basically, I was a recruiters dream. My decision was already made, and I even possessed a two-year degree, so my options were open. The sole reason I chose the reserves was the fact that I still needed to complete my education. I never really considered the Active Component. It's really hard to explain why I woke up that one morning and decided to join the Army. Within my immediate and extended family, only one person was serving in the military, a distant cousin, and in the Navy at that! In "the Early 90's," the military was still down-sizing, so as far as bonuses and incentives being offered at the time, I received none. As the old timers say, "I Joined for God and Country". The only benefit I received was the GI Bill, which later provided me a career I could never imagine.

I graduated from high school in 1990 and immediately attended community college in the fall to pursue an associate's degree as a radiological x-ray technician. This quickly panned out! Partying, surfing, and work, led me to change my career path and just pursue a normal AA degree program. If it were not for my parents pushing me to finish those two years, I would not be where I am at today, a commissioned officer. My first two years of college coincided with the

Gulf War. I was engulfed in the coverage of the war. This war instilled in me that I want to be a Soldier. However, I kept this to myself. If I had mentioned this to my friends, they would have thought I was crazy. As the years passed, my career was looking bleak. Then one morning in February 1993, my life was about to change forever.

My military career has been quite unusual compared to most Soldiers I have served with. When I first enlisted, I chose 88N (Transportation Management Coordinator) due to the fact I was color blind and could not select a combat MOS. I chose something that might help me in the civilian world. So 88N it was! My first two years in the Army was as a Reservist assigned to a Movement Control Team. The next three years as an 88N on Active Duty assigned to A Co, USAAMC, Fort Knox, KY. My job was as a household goods counselor and movement control team member. In February 1998, I decided to transfer to the AGR Program and was reclassified as a 75H (Personnel Admin NCO). During the next seven years I attained the rank of SSG (P) and was assigned to a Heavy Boat Company as the Company Admin NCO and to HRC-STL as an AGR career manager responsible for 450 78V/79Rs. While assigned to HRC-STL, I was going through what some people might call a "career crisis." I had just finished my Bachelor's Degree in Business Management from Columbia College and was getting extremely bored with my job and MOS. At this time, I talked it over with my wife and decided to submit my OBC packet for a direct commission in the AGR program. I even chose the Chemical field as a chance to broaden my range of military experience. Immediately after completing Chemical OBC, I was assigned as an Assistant Operations Officer at the 457th Chemical Battalion in Greenville, South Carolina. So far, the Chemical branch has been very good to me and has instilled that true loyalty/honor I originally felt when I first enlisted.

The wide array of job assignments has kept my attitude about the Army elevated. Over the last four years, I have been fortunate enough to complete my Bachelor's Degree and also complete numerous certificate programs ranging from Human Resource Management to MBA courses to Project Management courses. I owe the US Army for what it has offered me and my family. I even volunteered my services for this deployment, believing I owed my leadership, experience, honor and loyalty to the Soldiers I serve with. Predominately, I believe this is why most Soldiers enjoy the US Army. The ability to connect with people of all races, religions and color in something we all collectively believe in, "Serving Our Great Nation".

Why I Serve

Written by: SPC Danny Wilson, HHD, 15th PSB

I serve because of all of the fathers, brothers, and sons that came before me. I serve because of the tears of Old Glory's eyes. I serve for all of those kids that look up at a person in uniform with a twinkle in their eye. I serve for mom, apple pie, and Chevrolet. I serve for those special souls that never came back. I serve because I believe truth and justice triumphs over all. I serve for the peace of mind of future generations. I serve because I am inspired by NCOs, past, present, and future. I serve because of that history teacher that saw something in me. I serve because I believe in my fellow man and woman. I serve because of the tears I saw in my grandmother's eyes when I left home for the first time. I serve because I was told that I look good in uniform. I serve because of that fateful day in September. I serve because I want to make a difference in the world. I serve because the present military is the best trained and motivated in the history of our country. I serve in honor of my uncle who was a veteran of WW2. I serve because I want to be a part of something greater than myself. I serve because of the lost souls in Arlington Cemetery. I serve because of the vision of The Founding Fathers. I serve because I want to touch new faces and learn new things from them. I serve for the right to wear the 1st Cav Patch. I serve for those who can't defend themselves. I serve because it is part of my destiny. I serve because I am an American. I serve because it is expected of me to do so. I serve because life is short and you only go around once. I serve because voting booths in Iraq are a beautiful thing to see. I serve because when I look up at the sky I see Tar Heel Blue. That's why I serve!

Why I Serve

Written by: SPC Lyndsai Sustaire 376th PSB DET 5

"War is an ugly thing but not the ugliest of things. It's the decayed and degraded state of mind that thinks nothing is worth fighting for that is far worse." I read this quote in my soldier's manual when I was in basic training only three and a half years ago. It is a quote that has been with me since.

It's not about glory, recognition, the people you have killed, or even the friends you have lost. Picture yourself laying on fresh cut grass, family and friends around you, the smell of a new spring, hot summer or the sight of fresh snow falling all around you. Reminders of why you are here.

I didn't enlist for bonuses, or college money. I enlisted to prove something not only to myself, but to those who doubt the unknown, and to those who fear what they do not understand. I joined so my little girl can have the same choices and opportunities I did, along with everyone else. I joined in remembrance of the sacrifices our fellow Soldiers have given along with the sacrifices our fore fathers gave to this country. Those sacrifices are what make this country great; it's a devotion that even in death still lingers. I enlisted remembering the wars we have fought and won, the people who died, and the people who lived to share the memories. I want to save the world even if it is only one person at a time. A president may lead our country, but it's Soldiers who fight to keep it, and I am proud to be one of the many.

For the last ten months I have trained, prepared, and traveled to Camp Taji, Iraq, where I worked as a 63B (Light Wheel Vehicle Mechanic). To some this is a job better left for men in a place better left for those who originated from here. For me, it's doing something I love and will hold with me forever. That is why I serve!

Why I Serve

The Guardian
Written by: 1LT Tracey Ferguson

On the eve of a cloudy day in the windy city of Chicago (Illinois), a young lady questions her very existence and why fate dealt her the hand that she is now forced to endure. Here is a mother to one, a sister to three, a friend to a few and a colleague to many, who has to explain why she has to go to war when she is uncertain herself. In a few days unbeknownst to her, her family will grow and she will become a guardian to thousands. As a guardian, one is responsible for being the custodian, the gate keeper, the protector, and essentially the warden of whatever or whomever it is that he or she is to shield from harm and danger. When she took the oath to "support and defend" as an Adjutant General (AG) Officer, little did she know that those words would be the very thing to drive her existence in the months to come.

As a member of the United States Military, she was expected to uphold the core values (as is each Soldier) of loyalty, duty, respect, selfless service, honor, integrity, and personal courage. However, as an officer, the standard for each of those categories is higher and there is no room for the inability to perform adequately in any area. Combine that with being in a hostile environment and the daunting task to defend is magnified. Not only must she now defend the values that the constitution and the U.S. government holds dear, but also the lives and well-being of her Soldiers. She is their guardian. It is her duty to do everything that is morally, ethically and legally possible to ensure that the Soldiers' best interest is always at the forefront and that they return in the condition that they left, if not better. Being a part of the AG branch, one that stresses the ability to provide selfless service, her assignment asks for even more. Yes, it is a branch that has been labeled as the "paper pushers" of the Army. However, this is a branch whose emphasis is personnel. The goal is to provide the ultimate combat service support. Members of this branch might not be the Soldiers that are constantly going outside the wire and getting directly in harm's way but they do affect how those Soldiers are able to function on a day to day basis. It is the members of the AG branch that are responsible for handling all the Human Resources related functions of the military. They are the ones who process all the paper work to generate orders, awards, promotions, evaluations, and even the handling of postal operations (during war time).

The true burden of the Corps did not come to the forefront until the Soldier's deployment. It was then that she was able to see the whole picture and put her role and that of the Corps into perspective. She was much more than just the guardian of her Soldiers; she had also become the guardian of morale. She was responsible for ensuring the safety and movement of what wartime Soldiers have penned as "morale in a box." It is amazing how the little things like the mail influence the morale of the Soldiers that constantly place their lives in harm's way. The smiles and tears that are generated when Soldiers receive mail from their loved ones are priceless. When Soldiers are not given the opportunity to generate those emotions because there is no mail, the level of their spirits would diminish and their minds would no longer be focused on their respective missions, but begin to wonder on things back home. Therefore, it became her duty to make sure that those items moved as freely and safely to and from Soldiers as possible.

She now understands that being a part of the AG Corps is a blessing. It is a branch that is Soldiers focused no matter what state the Army is. There is no other branch that has such a strong influence on the daily interaction and state of mind of its Soldiers. It is a thankless assignment with few, if any, moments of praise. There are no Thank You's given or special recognitions when the Soldiers are able to receive their mail or send mail out to their families. It is something that is expected, and they believe they are entitled to. Your moments of gratification are self-imposed because you know that the fifteen plus hours that you and your Soldiers put in helped a family communicate with their loved one. However, the moments of blame are plentiful when the Soldier's precious cargo is delayed or does not arrive. It is in these moments that we are blamed for the lowering of morale and the lack of focus amongst the Soldiers. The command teams are not concerned with how the mail is routed or about delays in the system. All they know is that their loved ones have mailed packages that they have yet to receive, or that the birthday card that they sent has yet to arrive to their loved one back home. They do not care about the federal regulations that govern those who provide and move "morale in a box." All they see is the end result and what they think they are entitled to. It then becomes our responsibility to help them to grasp the underlying operation of the service that we provide, which they continuously question upon delays. It is not an easy task to have the power to influence so many by the completion of your job, yet be forced to do that job under the watchful eye of your fellow Soldiers and the federal government.

At the conclusion of her 16 months of deployment the mother, sister, friend and colleague will be able to return to her home with the majestic view of the Rocky Mountains. Her daughter will ask, "Mom, what was your role in the war and why did you have to go?" She can proudly say that she was sent to be a guardian. She will be able to rattle off stories of how she and her Soldiers braved the harsh conditions in their environment to provide the Soldiers the little things that we take for granted, homemade cards, notes of love and appreciation, care packages from loved ones and strangers alike. She did not have the glorifying task of going outside the wire and partaking in convoys. Instead as an officer, in general, she was the guardian of the lives of her Soldiers; however, as an AG officer, she became the guardian of much more. She was given the relentless task of maintaining the morale and strength of one of the greatest military elements in the world – The United States Army.

Why I Serve

Inspiration

Written by: 1LT Estela A. Oquendo

It is very common for little girls to play dress up as they are growing up; scrounging through their mother's dresses, trying on colorful make-up, and trying to walk in high heels. As a child, I could remember heading to the other side of the closet. The side where all the Army Battle Dress Uniforms (BDU's) were hanging and where the black spit shinned boots were neatly positioned. That side of the closet belonged to my father, who serve as an enlisted Soldier in the branch of Field Artillery. I never knew or understood at that time what being a Soldier entailed or even the United States Army's role in our country. All I could see as a child was a hard-working, dedicated, and compassionate man, who wore this uniform. My father started his daily work routine before anyone else in the house was out of bed, and he was usually home by dinner time. What he did throughout the day? I had no idea, but I could recall the family organization days, the Field Artillery shows, or just visiting my father at work as a slight insight to what his and the other Soldier's daily routine consisted of. Even at a young age and through my observations, the respect and discipline instilled in my father and the Soldiers was evident. These characteristics caught my attention and increased my curiosity of wanting to know more about the United States Army.

The opportunity to wear the uniform was available to me, both with the Junior Reserve Officer Training Corps (JROTC) in high school, and I was given the opportunity to continue my education through the Reserves Officer Training Corps (ROTC) as I attended college. Soon after I began my sophomore year, my emotions and commitment to the ROTC program and the military began to diminish, as my family was faced with a difficult challenge. My father, that Soldier in my life and my true hero, slowly weakened and became frail due to a battle with pancreatic cancer. As my father deteriorated, my dream to follow in his footsteps and become a Soldier decreased. I reflected on the numerous weeks he spent out in the field, his long tours away from home, or even the unspoken experience he endured while serving in Desert Storm. At that point, it didn't matter; I lost my father as well as my enthusiastic spirit.

After a couple of years of trying to put my life back in order, I knew that there was an indescribable vacancy to be filled. One day I was approached by a Soldier fresh out of Basic Training on hometown recruiting duty. Eagerly, I began the process of finally becoming a Soldier in the United States Army. Soon after arriving at Basic Training, the nation was faced with the horrendous tragedy of September 11, 2001. At this point, it was difficult to recognize if the willingness to be trained and the commitment to duty was a precedence in all of the trainees' lives. As I completed my training, the realization and understanding of why my father served in the military became evident and gave me the inspiration to continue to progress. I also now realized the importance of the military and its role for the United States and for the rest of the world. I willingly took the necessary measures to exceed, wanting to develop myself as a person and as a Soldier. Through mixed emotions, I felt as though I needed to protect my family and give support to all affected by this unfortunate incident. The extent of the September 11, 2001 tragedy gave a new meaning as to why it was important to defend our freedom as well as the country. As I began to appreciate what many service members have done before me, I realized that their accomplishments did not come easy. In order to successfully accomplish a mission, sacrifices need to be made, and adversities must be endured by both the service members and their families. The recognition of maintaining and securing the nations freedom was made real as war was declared.

Being an Army Reservist, I waited in anticipation to be called to duty, apprehensively waiting to see where and how I was going to serve my country. As Operation Enduring Freedom and Operation Iraqi Freedom were initiated, I began my civilian career as an educator in the public school system. The ability to work as an educator to America's future and at the same time fulfill my commitment of being a Warrior has provided me the opportunity to identify my eminence as a person. I have learned to separate both careers as well as identify specific traits each one provides in developing my character. From each practice I build upon my strengths and focus on improving my weaknesses. As I was called upon to lead a postal detachment in Operation Iraqi Freedom, I was well aware that the experience would be a test of the values I hold. As an educator, new students arrive every school year. Without knowing their background, there is an expectation and eagerness to establish a learning environment where there is progression and a sense of cohesiveness amongst the group. Given that the majority of the postal unit is cross-leveled, expectations as the platoon leader were similar. There was a differentiation of emotions knowing that the task at hand was vital to the lives of many. We were on our way to war and now my goal was not cultivating the minds of young children but leading Soldiers, successfully accomplishing the mission, and returning home safely to our families is now the priority.

As a junior officer, I have been provided with the opportunity to develop my Soldiers as well as myself. In taking on the mission with little knowledge and training as a postal officer, I looked to enhance my effectiveness as postal platoon leader. In the military there is a wealth of knowledge and experience within the service members who have endured the rigorous training and who have worked in similar if not the same positions. The training opportunities allow officers and enlisted to become proficient in their distinctive career field. I take each experience as a learning block to build up my character by analyzing every aspect related to the situation and my actions as an individual. I am cognizant of the fact that the decisions I now make affect more than just me and my family. My platoon's performance during this tour influences the lives of many personnel to include service members, civilians, and family members, something that is fundamental to our purpose and function in this battle.

When presented with the question of why I serve in the United States Army, I reflected on different aspects of my life. Growing up I had a role model, my father, who in my eyes was the epitome of a Soldier. He guided and influenced my decision of joining the military in order to continue a legacy which he set the path for. As I continue my journey of serving in the military, I understand the importance of training, continuous development, setting standards and improving deficiencies. It is important to know your job in detail and be cognizant of experiences that can foster individual ideals and ethics. As a leader, I need to be able to identify the differences in leadership styles as well as the evolution of the military and its service members which impact their values and morals. How people can influence others depends on how effectively each person sticks to their beliefs. I am proud to say that I have a role model that provided me with the inspiration to become a better person and has given me the motivation to lead the life I have chosen. I serve to be a role model and to leave a lasting impression on those I come across, the same as my father did.

Why I Serve

Written by: PFC Samantha Montemayor C-Det 15th PSB

There are many reasons why a person would want to join the military: to get away from home, to see the world, to go to college (cheaply), or to defend freedom and our way of life.

When I was asked why I joined, however, the answer may be quite shocking. I was lazy. In the Army someone else decides what I do, were I go and what I wear. I don't have to make decisions for myself. I put my laziness and complacency to work, I let someone else make my decisions, and I was going with the flow. I adapted easily with little complaints and I became a drone. Then reality set in, a sense of life if you will. So much has happened since I joined, and I don't regret a single minute of it. I'm not so lazy, now. I stand up, make my voice, along with my desires and my intentions heard. Now I am willing to work for what I want, and I finally know what I want.

When a senior NCO tells you, you're successful, that's when you realize you're in for a career. Nothing is a hand out in life. It's all about what you make in it. I know what we are doing over here is making a difference, and it makes me proud. I'm not married. I don't have any kids. My family is very supportive, and my nieces and nephews are proud of me. There is a personal pride in what I do. I want to be here and do my job to the best of my ability.

I had always dreamed of getting out of the little town of Portland, Texas. I always thought that I was better than what that town could offer, but I was always afraid of the unknown. Finally, I got up the courage to get away, and just join the Army; I would not change a second from the minute I stepped on that bus at Ft. Jackson until this moment. I have made many friends and lost touch with many, yet there faces and names are still with me. There are also the values and a sense of duty and responsibility to this country that I was raised with.

"Never walk away from someone who needs your help." Proverbs 3:27

I love everything about the military from meeting people from all walks of life to being a role model for the younger kids in my family. The reason I serve is because that I felt it was an honor to accept the call at a time of war. I serve knowing that I put everything aside for other people's lives and for mine, yet also knowing that my family can count on me to never give up on my beliefs. This is where I belong, serving my country.

Why I Serve

Written by: SPC Marcus Harvey C-Det 15th PSB

We are four years into the Iraqi war, 63 percent of Americans currently oppose the war, and over 3000 troops have died since 2001 in support of GWOT. Yet despite all of these obstacles, over 153,000 troops put on their uniform, and serve with honor every day in Iraq.

Why? The reason that we serve came to life before the GI bill, before any of our family served in the military, before 9/11, and before the first soldier lost his life.

The reason why we serve, Americans, Iraqis, and all other coalition forces, is because we all believe in the 200 year old words of Thomas Jefferson, that all "are endowed by their Creator with certain unalienable Rights, that among these are life, liberty, and the pursuit of Happiness…" to discover where were are, we must look to our past to see where we came from. Not everyone has family who served, or were so touched by 9/11 that they rushed to the nearest recruiting office, but each soldier, no matter what country, believes in these words.

It is the "soldier's burden" to ensure that children like these Iraqi orphans wake up in a place where there are those who will protect their life.

We serve because the oppression that was used by men such as Saddam and Hitler robs people of their liberty, which is morally wrong.

We serve because everyone has the right to pursue happiness, even if it comes in the form of a purple finger.

How do we win the war against terror and the other wars that are surely to come? The same way we have done for the past 200 years, by fighting not just because of events that have taken place, but by fighting for the belief that we all deserve life, liberty, and the pursuit of happiness.

Why I Serve

Written by: 1LT Chad M. Stormoen 2/175th AG Company (Postal)

Dear Sarah,

All of the officers in the 15th Personnel Services Battalion have been asked to write an essay for the Journal of the Adjutant General's Corps Regimental Association called the 1775. If you look at my green Class A uniform coat that is hanging in the bedroom closet, you will see the Adjutant General's Corps Regimental Insignia above the pocket with a red, white, and blue shield that has "1775" on the top and "Defend and Serve" on the bottom. That is the year that the Adjutant General's Corps was established. The essay that I am to write must address the reasons why I serve our nation in the United States Army as a commission officer and the tasks that I have been charged with during my tour in Tallil, Iraq, in support of Operation Iraqi Freedom, 2006-2008 rotation. You may be wondering why I am telling you all of this in this letter. Even though I have worked in an Army Post Office for the last six months, I have not been very good at writing and sending you letters. I do not have the excuse of not getting the chance to get to a mailbox. I decided to format my essay as a letter to you to help you understand better what I do and why I do it.

To best explain why I serve in the military, it is best that I start from the beginning of my career. While still in high school I was able to join and rehearse with the 132d Wisconsin Army National Guard Band prior to attending Army Basic Training at Fort Jackson, South Carolina. I still remember the sense of pride I felt when I wore the Battle Dress Uniform with my spit shined boots and ran into some of my classmates on my way to or from drills. Three days after graduating from high school, I shipped off to Basic Training. We met in college, fell in love, and eloped to Las Vegas to get married. Although we were young, I still believe we were pretty mature at that time and I remember having the discussion with you about my military commitment prior to getting married. You told me then that you supported my service to my country just as you support me now. After graduating from college with a Music Education degree, I decided to join the Active Duty Army as a musician; the same MOS I carried in the Wisconsin National Guard. I know you were not completely convinced that it was a good idea at the time, but once again you supported my decision to serve in the United States Army.

After two years of my Active Duty Army enlistment, I decided that the best thing I could do for my family, career, and county was to try to make a difference in the Army by commissioning through the ROTC Green to Gold Program. The ROTC program at Augusta State University gave me many practical leadership skills to prepare myself for a career as an Adjutant General's Corps officer. The Adjutant General's Corps was my first branch choice. I chose this field because I believe that skills gained from strength management, personnel actions, and other administrative duties can be of benefit to a successful career in education and educational leadership. During the last year and a half that I spent in ROTC, I attended classes at night and earned a Master's Degree in Education from Troy University at the Augusta Site which may help me with future endeavors in the Army or in our nation's public schools.

Once again you were supportive of the military lifestyle and me, this time as an officer's wife, as we drove cross-country to my first duty station at Fort Lewis, Washington. My first job was as the S2/Assistant S3 at the 22d Personnel Services Battalion, and when our battalion received deployment orders I requested the leadership challenge of becoming a postal platoon leader. The postal platoon leader job is the only platoon leader job available in the Adjutant General's Corps.

The 2/175th Adjutant General Company (Postal) was the last deploying element from the battalion. It was a great challenge to deploy without the support of the company or battalion headquarters which deployed months before we did. My platoon sergeant and I worked many hours together to ensure our platoon was technically and tactically prepared to go to war. Where others gave in to temptation to do the bare minimum so they could go home early, we exploited every opportunity to maximize the resources and time available to us for training. One training event that I will never forget was an underwater HMMWV rollover training cage. The training was not mandatory, but it should have been. The cage was designed because many Soldiers have drowned in HMMWVs in OIF because the poorly built Iraqi canal dikes that they drove on collapsed. My Soldiers were strapped in a mock HMMWV and were plunged into the pool water upside down, front over end. They then had to unbuckle their seat belt while sitting upside down under the water and navigate out of alternate locked doors while blindfolded. I remember the looks on some of the Soldiers' faces when they watched to demonstration. I saw fear in a few sets of eyes. It probably did not help that I was a first time NO GO in one of the tasks, as I was the first person to go through the chute. However, I was able to try it again and was successful at dismounting from the vehicle. The platoon had 100% participation in the training and all but one Soldier successfully received a GO in each of the underwater scenarios after multiple chances to do it correctly. I am confident that the Soldiers in my platoon have been given the best chance of survival if they would ever be flipped into a waterway while on a convoy on the way to deliver postal service to outlying forward operation bases. We did this because it was the right thing to do, not because anybody was tracking us or giving us praise or recognition.

June 4th, 2007, will mark six months of having boots on ground in the Middle East for the postal platoon I lead. During that time I have supervised 17 Soldiers who serviced thousands of postal patrons, moved millions of pounds of mail, and sold over half a million dollars in postage and money orders throughout four different post offices in Cedar II, LSA Adder, Camp Bucca, and Basrah which are all located in the Southern part of Iraq. I have been responsible for auditing both fixed and flexible postal finance accounts held by the Custodian of Postal Effects (COPE). Along with AAFEES, I was able to oversee the inclusion of Eagle Cash, a plastic cash card similar to a gift card, as an alternate way to purchase postage. Now Soldiers do not need to worry about carrying large amounts of cash to send heavy footlockers home which can sometimes cost over $100 depending on the types of services selected. Although official mail is not an Army Post Office responsibility, I recently was able to successfully mediate the differences between two senior officers concerning the proper handling of casualty mail and notification. Other than specific postal related issues that arise from time to time, I feel that I deal with the same issues that any other platoon leader would deal with to include supply, Soldiers' family issues, and training, just to name just a few. The postal mission that we are on orders for is important to the morale of the Soldiers who fight and the Soldiers that support the war fighters. The Army Post Office is one of three different ways of communicating between home and Iraq for Soldiers deployed to Iraq. The services offered through the Army Post Office are also the most cost effective way for Soldiers to receive comfort items from home.

I believe that I can make a difference in the Army as an institution and in the lives of Soldiers, no matter what the struggles may be. Of course, this is going to be tough on our relationship in the years ahead if I try to make a career and retire from the Army. It is hard to believe that I have over ten years of military service to our country. My appreciation for your support over the years is abounding. Thank you for being the best officer's wife a gentlemen could ask for. I miss you and I pray that God will keep us both safe and allow us to be together again sooner than later. I will do my best to try to write and send you more letters through the LSA Adder Army Post Office, starting with this one.

Love Always,
Chad

Deploying to Iraq

The Life and Times of a Deployed Reserve Officer
Written by: CPT Nicole Bires Commander, Detachment 4, 847th Personnel Services Battalion (Postal)

"So Captain, what exactly is your deployment history?" That is how my command of Detachment 4, 847th Personnel Services Battalion began in February of 2006. I had joined a Reserve unit from the Individual Ready Reserve in October 2005 in order to deploy-I got my wish a little bit faster than expected. I went from a Postal Company command in Chicago, Illinois to a Personnel Services Battalion based out of Fort Snelling, MN. From the blustery winter day, my weeks and months were filled with command courses, inventories, weapons ranges, and several meet and greets with my Soldiers. I had to leave my job as a Police Officer to take command of 50 Soldiers that I had never laid eyes on before. Such is the life of a Commander in today's modern Army Reserve.

Moving forward to July 2006, I finally got all of my Soldiers and NCOs in the same place at the same time. It was absolutely amazing to see this group of Soldiers from Illinois, Minnesota, Wisconsin, Ohio, and Indiana start to form bonds they would need to help each other through our mobilization and deployment to Iraq. In just a few days, we would all begin our intense training at Fort McCoy, WI.

Upon arrival to Fort McCoy, I found out that we would spend 2 ½ months preparing to deploy to one of the most hostile places in the world. Little did I know at the time, but my Detachment would spend most of those months training ourselves. We had the unfortunate luck to arrive at Fort McCoy during one of the busiest training times. Because of this, several units would vie for the same training time after time. The training that we did receive from the 85th Training Division at Fort McCoy was outstanding. Most of the instructors were NCOs and Soldiers that had just returned from Iraq or Afghanistan. That meant that they were able to teach us the latest in the Tactics, Techniques, and Procedures (TTPs) of enemies working in those theaters of operation. It was a relief that we were leaning the best tactics available at the time. We had classes and exercises that covered urban operations, ground convoys, combatives, reflexive fire techniques, base defense operations, improvise explosive device detection, and unexploded ordnance detection/procedures. One of the only things we didn't train on was how to do our jobs. We knew that we would have to learn what to do and how to do it using on-the-job training.

During several joint training exercises, I was able to meet several other commanders that I would run into and meet throughout my time in Iraq. I had the distinct pleasure to attend a three-day leadership seminar with a rather outgoing and charismatic bunch of Engineers from the Alabama National Guard (1169 ENG). They always had a kind word and a bit of advice to give, regardless of the situation. It remained the same when I was able to link up with them in Iraq. Even though they had some of the toughest jobs in country, they were always happy to see me. It was refreshing to have a group of fellow "part-timers" to laugh and joke around with. One of the biggest jokes was the Reserve/National Guard motto "two days a month, two weeks a year" that is a mainstay of the recruiting world. We would always answer with a hearty "YEAH RIGHT!" But in all seriousness, we all knew we had important jobs to do, and were proud to be allowed to do them with the greatest Army in the world.

After learning everything we could from our mobilization training, it was time to depart the greatest country in the world to begin our jobs of defending it. It was now time to put our training to use by supporting the war-fighters with our mission of flight operations. We departed the Midwest on a cold day in October for a land that only a few of us had ever seen before a land that was not known to welcome Americans with open arms.

Our arrival in Iraq was met with a flurry of activity. There was much to do prior to 4/847th PSB taking over the R5 (Reception, Replacements, Return-to-Duty, Rest and Recuperation, and Redeployment) and DTAS (Deployed Theater Accountability System) missions at Baghdad International Airport's (BIAP) Joint Reception Center, and on Camp Liberty from B Detachment, 502d Personnel Battalion. Once my Soldiers were settled, they were ready to get to work.

One of the first orders of business was to support the 15th Personnel Services Battalion with personnel that would allow them to complete various missions. I selected seventeen soldiers to support the 15th PSB in their automation, maintenance, personnel, casualty, and headquarters sections. It was tough to give up quality Soldiers to a unit that I barely knew, but I understood that those critical missions could not be as easily completed without the hard working and dedicated Blue Devil Soldiers of DET 4.

The next several days were spent learning the jobs (the right seat) that would be required of us for the next year. When the Soldiers were confident in their abilities to account for over 33,000 Soldiers and Civilians, manifest passengers, escort personnel to and from helicopters, and deal with an array of customers with varying levels of emotions, we moved over to the left seat and started performing the jobs that would be completed 24 hours a day, seven days a week for at least one year. On November 17, 2006, we officially took over the mission from the 502d PSB. It was now our show to run. There was no looking back, and no one to help out with any leftover questions. The 502d PSB redeployed and just like that, we were responsible for accounting and moving thousands of personnel around the theater.

The Soldiers seemed to jump into their missions with both feet. The NCOICs at BIAP saw that many changes needed to be made to enhance passenger flow and provide better customer service. They ended up making some drastic changes that ultimately improved our customer wait time for movement on helicopters to outlying forward operating bases (FOBs). Some changes were made that had been planned for more than one rotation. Construction on a new hard stand passenger terminal was started. The construction took away our staging area for flights, so an additional temporary staging area was built. In the long run, the terminal will be an improved place to work, but I'm not sure it will be done before my unit leaves theater. My Soldiers also get to work with different components. As BIAP is actually an U.S. Air Force asset, we will have worked with four different rotations of Air Force personnel by the time our deployment ends (they are on four month tours). There is also representation from the U.S. Navy, the Royal Air Force, and civilian contractors stationed at the passenger terminal. It's a great learning experience to see how different services and countries operate. It is not something that the average Soldier gets to see every day.

It really is amazing to see how well the Soldiers learn and adapt to their environment. The peculiar thing is, if you take away the mortars, small arms fire, and the fact that we walk around with weapons all day, the Victory Base Complex is very garrison-like. There are many rules and regulations that you would find at any post in the military. Common customs and courtesies are still executed and standards are still in place. Just because we are in an environment that is inherently dangerous, doesn't mean that our Soldiers are allowed to do whatever they want. Seat belts and helmet are still worn in vehicles, full uniforms are worn to and from the latrines on the living pad, and rooms are inspected at least twice a month. There are mandatory safety days, urinalysis testing, and refresher training on prevention of sexual harassment, consideration of others, and escalation of force. We have monthly OPD and NCOPD sessions to enhance the learning of leadership within the Battalion and yes, even papers to write about our experiences in Iraq. All of these tasks, conditions, and standards do actually have a purpose. I may have a hard time finding the purpose sometimes, but I know it is for the health and well-being and professional growth of all Soldiers, NCOs and Officers within the 15th PSB.

Overall, I would have to say that my deployment has been a huge growing experience for me. I have learned so much about myself personally and professionally. I continue to learn daily how to lead the finest Soldiers in the Army Reserves. I know that not all days are easy, just like I know that the best days must end sometime. I only hope that the best days outnumber those that are difficult. I made exactly two promises to my Soldiers and their family members back on a cold winter Wisconsin day. The first was that we would execute our missions to the best of our abilities. The second promise was that I would do everything in my power to get each one of them home to their husbands or wives; sons or daughters; fathers and mothers. On the first, I can say that we execute our mission as best as we can on a daily basis. We provide outstanding customer service and get people out of the country on rest and recuperation leave-one of the most important breaks for Soldiers that may have seen way too much while here. As for the second, I can only pray every day that I am doing what is right for my Soldiers. I know that I can't control everything that the world throws at them, but I do my best to ensure that they will have the opportunity to get on a plane at the end of our tour with their heads held high and pride in their hearts from a job well done.

"DET 4. READY FOR WAR!"

Deploying to Iraq

PSB- The Last Stand

Written by: CPT Rodney Davis 15th Personnel Services Battalion

The date was the 10th of October 2006; we knew months earlier that we would be deploying, but no one actually believed the day would come. As we staged in Raider gym, awaiting the arrival of the bus transportation to the airfield, a lot of thoughts went through my head.

Being a veteran of OIF II, the first and foremost thought was to ensure that all of my soldiers made it home. You see, I never imagined as a Field Artillery officer that I would be conducting combat operations as an infantryman, managing three maneuver platoons throughout the Al Rasheed District of Baghdad.

Now, I am on my second tour to Iraq with a different group of soldiers and a totally different mission. My soldiers and I arrived in theater with one thing on our minds, accomplishment of the mission. Alpha Detachment, 15th Personnel Services Battalion, embarked on a mission to provide personnel services to over 60% or two-thirds of the Baghdad area of responsibility. As the Adjutant General Human Resources command transforms into the next echelon of personnel services, the day will come when we, the PSBs, will no longer be a part of rebuilding Iraq. We unfortunately will never know when that day will come. For Soldiers of Alpha Detachment 15th Personnel Services Battalion, the beginning started on 10 October 2006 when we all boarded the plane bound for Camp Buehring, Kuwait. Sensitive items all accounted for and the Soldiers of Alpha Detachment bid farewell to Fort Hood and boarded the plane.

Eighteen hours later we were all greeted by 110 plus heat index. Two weeks at Camp Buehring was an eye opener for the Gators of Alpha Detachment. For some soldiers, this was their first deployment and it was the first time they have been away from home. The soldiers really enjoyed the training that they received in Kuwait. The training ranged from IED classes to a 3-day FTX to Convoy Live Fire Exercises with one of the deploying Infantry units.

We also linked up with some of detachment soldiers that deployed two weeks prior to the Battalion. Those soldiers led by SGT Chapell Williams and SGT Willie Crook, were selected to provide personnel service support for the war fighters in Ar Ramadi and Al Asad, Iraq respectively. Along with personnel service, we also provided a Casualty Liaison team and an R5 team for their perspective areas.

After the two weeks spent training in Camp Buehring it was time to board another plan headed to Baghdad, Iraq. Soldiers were packed in the C130 Hercules and whisked away in the middle of the night. The mission was finally here. It would be time for the Soldiers of Alpha Detachment to perform their wartime missions. All the training and preparation came down to this point. The briefings, lectures, and stories told by myself and other OIF II veterans were becoming a reality. The plane touched down around 0230 on the 29th of October 2006.

Landing at Baghdad International Airport was a somber moment for some. For me, I couldn't believe that I was back in Iraq 18 months later. It brought back memories of the Soldiers my Battery lost here the last time. My job this time was to fight complacency rather than the enemy whom is still living outside the walls of Camp Liberty Iraq. Keeping the Soldiers gainfully employed and focused on the mission at hand.

The RIP/TOA with our counterparts was seamless and fluent. Kudos to the 502d PSB for maintaining the standard of success that was left by other Personnel Services Battalions, to include this one.

The date was the 11th November 2006, exactly one month to the day of deployment and the Transfer of Authority was complete. The 15th Personnel Services Battalion took charge of the Baghdad Area of Responsibility. Alpha Detachment, led by CPT Rodney L. Davis spearheaded the way in taking over responsibility of the bulk of the mission by providing services to well over 50,000 Soldiers, Sailors, Airmen, Marines, Corpsmen, Department of Defense Civilians and Independent Government Contractors. Our Area of Responsibility consisted of Multi-National Division-Baghdad, Multi-National Corps-Iraq, Multi-National Forces-Iraq, and Multi-National Forces-West. Our missions challenge was to provide personnel services along with Common Access Card (CAC) issuance, CAC Personnel Identification Number Resets, a Casualty Liaison Team, and an abbreviated R5 mission, encompassing the largest Personnel Service Support (PSS) mission in the battalion. We also detached Soldiers in support of two of the five brigades within the 1st Cavalry Division. Eight Soldiers, four each in the 1st and 2nd Brigade Combat Teams, provide personnel support to their brigades and battalions that are organic and nonorganic to the Division.

In the first five months of deployment, Alpha Detachment produced over 12,500 Common Access Cards (CACs), performed over 8,000 PIN Resets, reviewed over 3,000 promotion packets, updated over 4400 Officer and Enlisted Record Briefs, initiated over 5000 personnel actions and 1500 Good Conduct Medal submissions, submitted over 250 Naturalization Packets, and made over 4000 identification tags. What a phenomenal effort on the part of the Soldiers. Of course, the detachment can't function without its Human Resource Technician, CW2 Melba Mosley. The professionalism and customer service this detachment has is truly a direct influence due to her dynamic work ethic and leadership.

The "Gators" continued to influence the battlefield by providing excellent and recognizable customer service throughout the Baghdad AOR. Alpha Detachment conducted monthly G1/S1 conferences to keep the brigade and battalion S1s informed of changes within the PSS arena. The NCOICs of each section contributed to the topics by providing trends that directly affect the Soldiers and their actions. Alpha Detachment was also tasked to coordinate and execute

Personnel Service Support rodeos, which allow Soldiers from the PSB to travel throughout the battle space conducting personnel service operations. Headed by SSG Medina, the Human Resources NCOIC, the team has provided a myriad of services to outlying Forward Operating Bases and Command Outposts that do not have a PSB in their foot print to support them. The team produces CACs, conducts PIN resets, and updates Soldiers Record of Emergency Data (DD93) and Servicemen's Group Life Insurance (SGLI). With the deployable CAC machine in hand, the team sometimes spent two or three days off their own FOB conducting operations. These missions have proven to be a vital asset to the war fighters because it limits their time on the road coming to us for service. The Soldiers really enjoy the experiences they gain from the mission; they also had first-hand knowledge of how other Soldiers were living in places that do not have the luxury that they do on Camp Liberty. The experience gave the Soldiers a new outlook on life and they learned not to take their accommodations for granted. There are far worst places we could have been instead of being here.

In April 2007, Alpha Detachment geared up for a new mission, which consisted of supporting Multi-National Division- Central, supporting 3rd Infantry Division as they moved into our AOR. The soldiers continued to impress me with their "Can Do" attitudes and willingness to succeed at any mission placed before them. As we transitioned into the 3rd quarter of FY07, we continued to improve the personnel services provided to our battle space by taking the services to our own 1st Cavalry Division or MND-B headquarters, providing them with DD93 and SGLI updates and answering any questions on the Family SGLI for dual military.

Alpha Detachment would not function without the personnel behind the scene. If the NCO's are the backbone, then the headquarters section is definitely the "heartbeat" of this detachment. Master Sergeant Scottie Jarrett, SPC Chantele Monares, and SPC Ann Jefferson perform the daily functions of the Detachment Sergeant, training clerk, and supply clerk. Behind the scenes of every great PSB, there is a dynamic HQ section that caters to the needs of their Soldiers. From conducting physical fitness training, sensitive items accountability, ordering supplies, updating records or processing leave packets, these three individuals deserve high recognition of this detachment. I could not ask for a better team.

Finally, as the Alpha Gators continue to Live the Legend of the 15th Personnel Services Battalion, we are grateful for the men and women that we support daily who are out on the battlefield risking life and limb in the name of democracy. Their courage and commitment to the people of Iraq are truly appreciated by the support arena, the Iraqi citizens, and the United States of America. On behalf of Alpha Detachment, 15th Personnel Services Battalion, I salute you, America's sons and daughters for the sacrifice that you have made to fight for freedom and justice throughout this land.

Deploying to Iraq

The Road to War

Written by: CPT Seth C. Graves 15th Personnel Services Battalion

The challenges in preparing Headquarters and Headquarters Detachment (HHD), 15th Personnel Services Battalion for its deployment to Iraq, in support of Operation Iraqi Freedom (OIF) 06-08 have been many. The Detachment's mission is to rapidly deploy the 15th PSB to designated contingency areas and facilitate personnel services detachments and postal companies to provide full spectrum personnel service support and postal support through sustainment and support operations and then redeploy. The Battalion knew of its deployment for OIF 06-08 approximately a year in advance, which would lead one to think that our unit had more than enough time to prepare for a deployment that could potentially last upwards of 16 months; that is not the case. HHD was particularly challenged with the fact that while we had to meet all training and readiness requirements of the Battalion's other subordinate units HHD was also responsible for providing daily support to those subordinate units.

In preparing for OIF 06-08, all Soldiers were required to train on over 80 individual, leader and collective tasks. The tasks ranged from performing first aid to culmination events, such as participating in convoy live fire exercises; no small task considering a large percentage of the Soldiers in HHD had no deployment experience and required large amounts of train up to execute many collective task, not to mention, our Soldiers provided continuous support to other units. The Battalion leadership saw the difficult road ahead that faced me and my First Sergeant and committed two training days per week where the Soldiers on staff would have dedicated time to train. Staff sections were required to provide on call assistance if needed and we would always allow staff elements to leave training in order to ensure the Battalion requirements were met, which was often. Another challenge was keeping Soldiers motivated; reminding them to execute many of the required deployment tasks, they are all Soldiers first and human resource specialists, or whatever their MOS might be, second.

Now that we had dedicated training time, my First Sergeant and I developed a comprehensive training plan, crucial to ensure we got the maximum training benefit for our Soldiers. We started by planning a Detachment field training exercise (FTX), which incorporated another Detachment within the Battalion. This maximized the use of our noncommission officers as trainers, which enabled realistic training and allowed us to focus more time on training events.

Many post resources were utilized, such as the convoy simulator and Engagement Skills Trainer 2000 (EST2000). The convoy simulator was extremely helpful due to the fact that we could start pushing our Soldiers through at 0530. By doing this, we were able to get Soldiers through multiple runs in different crew positions while having them report to their staff sections by 0900 on days that weren't one of our two dedicated training days. The convoy simulator provided great dividends in preparing our Soldiers for convoy live fire exercises. HHD also heavily used the EST2000. EST2000 not only provided Soldiers great confidence in target acquisition, it also further improved marksmanship skills, which was evident at numerous qualification ranges. The post-run M16 qualification range was also heavily used, which enabled the unit's Soldiers to qualify on their personal weapons without the personnel drain associated with planning, preparing and running a range.

With many required training tasks now under their belt, the Detachment's Soldiers participated in a Battalion FTX. The FTX not only provided a great opportunity to complete required training tasks, it also allowed the Soldiers on staff to exercise systems and procedures that would be in place once in theater, such as JPERSTAT and casualty reporting, manning of an Emergency Operations Cell, etc.

The Soldiers were run through numerous STX lanes, which ran the gamut from donning NBC protective equipment to convoy operations.

Upon completion of the Battalion FTX (approximately deployment - 90 days), the Soldiers of HHD were close to being complete with required training tasks and ready for the convoy live fire exercise. The convoy live fire exercise enabled the Soldiers to put to use many of the tasks they had trained on over the previous months. Soldiers built confidence in their ability to engage and put metal on targets from a traveling convoy with personal and crew served weapons. Soldiers also trained on techniques for identifying and reacting to improvised explosive devices as well as vehicle recovery and MEDEVAC procedures.

With the majority of the unit's required training now complete, we could focus entirely on Soldier readiness requirements; as well as sustainment training as opportunities arose. All Soldier and Family readiness requirements were tracked weekly on the Division's Battle War Winning Report (BWWR) The BWWR was a comprehensive list consisting of over 100 metrics. Everything concerning Soldier and Family readiness was included such as unit immunization statistics, hearing and vision testing, DD93, and SGLV, wills and power of attorney, and Family readiness data, just to name a few. Commanding a unit that was mainly comprised of the Battalion's staff definitely had its advantages. The S1 section provided assistance in updating DD93s and SGLVs, the legal clerk assisted in processing requests for wills and powers of attorney, S3 provided the unit with assistance in coordinating for immunization rodeos, and the Battalion's NBC NCOIC was able to conduct PATS testing and inspection of NBC gear and equipment. The staff truly pulled together and provided a jump start for the unit prior to going through Soldier readiness processing. The Detachment was fortunate to have a family readiness group (FRG) leader in place, who had led the Detachment's FRG during the unit's first deployment for OIF II. The Soldiers and families of HHD were now trained, ready and informed for OIF 06-08.

Shortly after arrival to Camp Liberty, Iraq, the 15th PSB conducted a transition of authority with the 502d PSB. The Soldiers of HHD were eager to dig in and accomplish the mission. In preparation for OIF 06-08, the staff's numbers swelled. By MTOE a PSB HHD is authorized 27 personnel and is not fully equipped to provide the proper level of support in a garrison environment, much less a deployed one. The 15th PSB doubled in size upon arrival in theater with the addition of two personnel service detachments, two postal company headquarters with a combined total of thirteen postal platoons. The increase in subordinate units and level of support required the Battalion's staff to reorganize. Much of the reorganization took place prior to deployment and some was still in place from the Battalion's previous deployment during OIF II. HHD's numbers peaked at 53 Soldiers, with 42 being assigned or attached and 11 being OPCON from the Battalion's subordinate units.

Unique to garrison operations, the Battalion stood up an Emergency Operations Cell (EOC), which was the center of gravity for daily operations. The EOC monitors command nets and is the conduit for information across the Battalion's battle space, which has Soldiers dispersed at 28 different locations over the Southern two thirds of Iraq. A S3 Air section was created to manage and track the numerous air mission requests that are mission essential to conduct movement throughout the Battalion's battle space. Just seven months into the deployment, over 600 air mission requests have been processed. A postal operations officer position was created to ensure dedicated support to the Battalion's postal elements. The S6 section doubled in size, receiving two Soldiers from one of the Battalion's Reserve Personnel Services Detachments; the Soldiers have been a great addition and come to us with an extensive civilian background in automations support, which has been essential in keeping the myriad of networks and automated personnel service systems fully mission capable. The S1 section received a Reserve Soldier as well, who has proved to be an invaluable asset when processing Reserve Soldier personnel actions. The battalion's maintenance section received 5 additional Soldiers to assist in keeping the Battalion's fleet and rolling stock, over 100 pieces of equipment, fully mission capable.

The Detachment's Soldiers worked and trained hard in preparation for their OIF 06-08 missions, ensuring the Detachment deployed with a T rating for its METL tasks, deployed with the proper equipment, and left ready and informed capable families behind to secure the home front. The Soldiers of HHD continue to provide first class support to a Battalion who provides personnel service support, R5, casualty reporting, personnel accountability and postal support to over 96,000 service members across the lower two thirds of Iraq. I'm truly proud to command such fine Soldiers; they continue to amaze myself and the leadership throughout the Battalion every day.

Deploying to Iraq –Leader Development Matters in Combat

Leader Development
Written by: LTC Angela M. Odom 15th Personnel Services Battalion

Team building in this combat environment is critical and extremely necessary. We established an effective leader development program for our geographically dispersed Personnel Services Battalion the week after our formal Transfer of Authority from the 502d Personnel Services Battalion. This was initially driven by the requirement to update our higher headquarters, and then, more importantly, by the need to ensure our leaders in 31 separate locations were providing the same level of professional Essential Personnel Services; Postal Operations, R5 Operations, and Casualty Tracking support.

The fact that we have Captains, Lieutenants, Staff Sergeants, and Privates communicating one on one with warfighters made it all the more pertinent that our leaders are well-trained, competent, and confident. Some deployed with the experience of multiple deployments and a myriad of Active and Reserve Component positions. Some deployed with outstanding relevant tactical training and absolutely no mission related training. All deployed with a willingness to learn and receive updates to their initial mission. Regardless. Our intent was always to set our Soldiers and leaders up for success.

The first 60 to 90 days of our deployment, we focused on establishing battalion-wide systems; mostly administrative in nature. We had to immediately focus on the thing that a professional organization routinely accomplishes. Our battalion deployed with four of its organic elements from Fort Hood, Texas and met the other four elements in theater. I arrived in theater after the 15th Personnel

Services Battalion's eight subordinate elements deployed and immediately conducted an Assumption of Command ceremony.

We determined that the intended results were to develop effective plans to achieve organization goals, to set clear priorities, to create alternate or contingency plans, to consistently meet deadlines and commitments on time, to consistently achieve mission tasks and objectives, and to provide a command and leader presence (right time/ right place).

We established a professional development training plan for NCOs and officers. We determined our assessment/ feedback would be from monthly OPD (AG related) by Commanders and Staff, by daily feedback and professional development, and by a daily 15th PSB Battle Update Brief. Additional feedback was received through the weekly 15th PSB Command and Staff meetings, written azimuth checks, and command visits.

Deployment Timeline

Our first six months in theater have been filled with great challenges, and I am proud to say that our leaders have addressed those challenges with ease and confidence. We have junior Soldiers who can provide detailed updates on casualties without blinking an eye, Sergeants who can manage inbound units of 400 personnel through APODs, Lieutenants who can communicate with senior officers from maneuver battalions who can provide short term postal augmentees, personnel technicians who coordinate with other technicians across the PSB's Area of Responsibility to complete Personnel Asset Inventories ahead of DA's schedule, and of course Commanders and First Sergeants who take care of Soldiers in the manner in which the Army demands.

Leader development matters in Iraq. We come with varying degrees of skills, knowledge and abilities. We come with a wide range of capabilities and concepts of what is appropriate and acceptable. It is imperative that we train all leaders consistently and professionally.

As the last PSB standing, we ae eager to set our Soldiers up for success as they will undoubtedly be assigned to a Brigade Combat Team or HRSC in the very near future.

Deploying to Iraq – The Fifth Detachment

Leader Development
Written by: CW3 Mark Fraley 5th Detachment, 376th Personnel Services Battalion

We began building a Team in April 2006, The 5th Detachment of the 376th Personnel Services Battalion from Long Beach, California was the first full Detachment from the PSB to be built from scratch. The organic Detachments 1, 2, and 3 had just returned from their tours in Iraq. Detachment 4 was a small Headquarters team stationed in Balad and ready to redeploy soon. We were made up of 25% new 376th PSB Soldiers, 25% volunteers from other units within the 63D Regional Readiness Command. We drew Soldiers from California and Arizona to fill the "Battle Roster" making this Detachment's background truly diverse.

In July 2006 we departed Long Beach, California for Ft McCoy, Wisconsin and two months of Pre-Deployment training. Any of you who have been to Ft. McCoy will probably attest to the beauty of Wisconsin and at the same time the utter distain you have for the Fort and its deplorable living conditions. Although it was the first time living in the "WWII Open Bay" buildings with no heat, air conditioning, or window screens for the majority of the Soldiers, they did not complain.

The training went well (if you were an 11B), and we managed to get out of Ft. McCoy with only minor injuries. I wondered why there wasn't any of the "mission related" training we had requested, even posed that question to our representative at the MUIC. Why didn't we have classes on how to make Common Access Cards (CAC), attend Casualty Liaison Team training, or get Postal qualified? "Not Enough Time" and "No Classes Available" were the answers. With all the down time (UTT) we had, the entire Detachment could have attended the F5 Postal Course. Isn't hind sight wonderful?

The lone arduous flight to Kuwait was exhausting for some and a chance to catch up on sleep for others. Once on the ground in Kuwait we began the trek to Camp Buehring for a week or two of training and acclimatization while waiting to be called forward. Day one was 127 degrees and windy. Day two 127 degrees and windy. Day three was 127 degrees.....well, you get the picture. Staring across the Kuwait landscape I likened it to the "Moon" and wondered how and why human beings would ever live here in the dirt? Oh I forgot; there isn't any sand in this desert! Its dirt! Dirt so fine that it eventually gets in to everything, no exceptions! Oh how I wish I had invented "canned air"! How could this place have sustained any life? I just saw a beetle run past my foot....What did the beetle live on? Could I live on beetles?

Finally, after three mission changes and countless emails and phone calls to theater we received our "call forward." The Commander and Detachment Sergeant had gone with the Advance Party to Balad, Iraq to get organized since this was to be our headquarters. About 10 hours before the main body departure time word came from the commander; "Mission Change!" We were going to Balad but weren't staying. The new and "final" mission change had us replacing a detachment from the 22d PSB in Camp Taji, Iraq. No one had ever heard of Taji and we all wondered where it was.

As 0400 hours came, we quickly loaded our gear on to the truck and climbed on the busses for the long ride to the Air Field. At Ali Al Salem, we palletized our bags and then waited to be called to our aircraft. The C-130 was packed and we were finally on our way. A couple of hours after take-off, we landed at the Balad Air Field and were taken to the PAX Terminal area to wait for our Air Lift to Taji. It was quite late when the two CH-47s landed to take us to Camp Taji. After loading all of the bags the crew chief of one of the Chinooks said he was out of weight and balance so we had to transfer 25% of the bags to the second aircraft, what a mess, doing all of that was pretty chaotic. The flight seemed to take an hour but it was only 15 minutes. Low and fast over the Iraqi countryside in complete blackout reminded me of my years as a Rotary Wing Aviator with the Combat Aviation Brigade. Upon touchdown we were greeted by the Detachment Commander and Detachment Sergeant who had already completed a recon of our new home for the next year. We had to make some quick decisions regarding our living quarters but managed to clear that hurdle and started settling in.

The building that would be our worksite for the next year was in need of some TLC but for the most part was "air-worthy" and ready to go. Tasks were assigned, offices were emptied and reorganized, communication lines were repaired and run, and we got to the task of opening a full service Personnel Services Support center. The Soldiers of Camp Taji were starting their annual rotation, so the lines for Common Access Cards issue, PIN Resets, Evaluation Reports and Record Brief Updates were long and so were the hours we kept.

We thought we were going to be working for the 376th PSB out of Balad, who was soon to rotate out and be replaced by the 22nd PSB, but soon found out we had been reassigned to the 502d PSB out of Camp Liberty, Baghdad who was soon to be replaced by the 15th PSB. We learned later that all the name changes and shuffling were due to battle space realignments taking place in theater and we just happen to be sitting on one of the battel space border lines. We ended up with our Battalion Headquarters at Camp Liberty, Baghdad which required us to travel by air for meetings, briefings, conferences, etc., often enough to make you start thinking about the possibilities of being shot down. Two of the flights I have taken have had to return fire and at least twice that have popped chaff and flairs while traveling between Taji and Baghdad. "Popping Chaff and Flairs" typically means you have been or are being acquired by some sort of Surface to Air Missile (SAM). We quickly set out to meet the Brigade's S-1s at Camp Taji and establish our working relationships. We designed and offered all of them classes in CAC Ops and PIN Resets, 1506 Calculations, CSB/REDUX, and other PSS functions to help get them ready for the upcoming Personnel Services Delivery Redesign (PSDR).

PSDR migrates the core of HR support to TOE BDE and BN S1 sections; provides additional personnel and equipment resources to the TOE S1; and provides new theater level TOE HR units for casualty reporting, postal operations, and R5 (reception, return to duty, rest and recuperation, replacement, and redeployment) Personnel Accountability Teams (PAT).

The BCTs ate the training up like starving children. We made a lot of money with that training. Since we are now inside 90 days of containerization for our redeployment and the AC has been involuntarily extended for 90 days, we have offered the CAC and PIN Reset training and certification, CSB/REDUX, and general Personnel Actions review to the BCTs again to ensure they are ready and trained up to take lead in all areas of S-1 as we depart theater this fall.

Deploying to Iraq - 15th PSB Soldier Earns Exceptional Award

Written by: 1LT Steven Druckenmiller

SGM Larry Strickland would have been proud. On Tuesday, July 17th, 2007 one of the 15th Personnel Services Battalion's finest Soldiers received the honor of a lifetime. CPL Brandon Wildeman, an NCO in the PSB's S1 Section, received the SGM Larry Strickland Medal in recognition of earning the Adjutant General Corps Regimental Association's Soldier of the Year Award. What made the awards ceremony so special is that SGM Larry Strickland's widow, CSM Debra L. Strickland, was on hand at the 15th PSB to personally bestow the Medal on CPL Wildeman. Additionally, the CSM for Multi-National Force-Iraq, CSM Marvin Hill, partook in the ceremony.

While serving in the rank of specialist, CPL Wildeman was nominated by the 15th PSB's senior NCO, CSM Timothy Livengood. When asked, CSM Livengood expressed his confidence in CPL Wildeman and Combat Service Support Soldiers in general:

CPL Wildeman was selected because he exemplifies the Army Values and possesses all the character traits that we want our Soldiers to have today.

Soldiers throughout Iraq are out on the streets conducting combat operations and are labeled heroes. I believe they are heroes and have a definitive impact on the mission here in Iraq. However, you don't need to be on the streets to have an impact on others as CPL Wildeman shows daily. CPL Wildeman takes great pride in the fact that he has a direct impact on 465 Soldiers and Airmen of the battalion through his work in S1. He is responsible for the daily JPERSTAT, picks up/delivers mail, and helps with awards and promotion boards when needed. All of these areas are combat multipliers through raised morale in our Soldiers and Airmen.

The SGM Larry Strickland medal was named for the Adjutant General's Corps top enlisted member, who worked in the G1 section of the Pentagon. SGM Strickland was one of the 125 victims of terrorism perpetuated by al-Qaeda on September 11th, 2001. The Executive Council of the Adjutant General Corps Regimental Association (AGCRA) responded by seeking ways to honor Adjutant General Corps' Soldiers lost in the War on Terror. LTG Timothy Maude and SGM Larry Strickland were both honored posthumously when the AGCRA approved the LTG Timothy Maude Medal (for the AG Officer of the Year) and the SGM Larry Strickland Medals (for AG NCO and Soldier of the Year).

The prestige of the award, coupled by the attendance of Iraq's senior CSM and CSM Strickland (currently the Installation Management Command [IMCOM] CSM) is a high point for the 15th PSB, and specifically CPL Wildeman.

The fact that CPT Richard Strong, the AG Officer of the Year and recipient of the LTG Maude Medal, is serving with the 15th PSB (and was on-hand for the ceremony), underscores the dedication of each Soldier in the PSB, and this forward-deployed unit as a whole, to living the AG motto, "Defend and Serve!" The 15th PSB will continue to train and employ highly-trained, motivated Adjutant General Troops, and this recent award highlights the Battalion's achievements further.

The Military Postal Service

Military Postal Operations
Written by: 1LT Migdalia Summerville

The origin of Army Postal Service dates back to 1856 when the first Field Post Office accompanied the British Expeditionary forces to Persia. The Field Post Office continued to be requisitioned by the Army from the Department of Posts to accompany Expeditionary Forces until the First World War. During the Second World War, volunteers from the Department of Posts and Telegraphs were drawn as combatants to run the Field Post Office. In 1947, when Indian troops were rushed to Jammu and Kashmir, a handful of personnel, who remained with the occupation forces in Japan, were sent to provide postal services to the troops. This group developed into the present Army Postal Service. Initially the organization was affiliated to the Army Services Corps, but in March 1972 the Army Service Corps was formed into a separate corps.

Military Postal Service is a service composed of the Military Postal Service Agency and the aggregate of Department of Defense commands, organizations, personnel, and facilities that provide postal services for authorized personnel and organizations outside the United States and its outlying areas. The mission of the Military Postal Service Agency is to achieve efficient and economical transportation and distribution of official and personal military mail throughout the Department of Defense, and effective and responsive operation of Military Post Offices. The function of the Military Postal Service Agency is to functionally manage the Military Postal Service by maintaining direct liaison with designated theater postal activities and the Military Departments to ensure the effective integration of military postal transportation and distribution processes, and the implementation of uniform Department of Defense postal practices and procedures worldwide. Also, the Military Postal Service Agency must maintain Joint Military Postal Activities at the United States Postal Service military mail gateways within the United States to provide liaison for Military Postal Service activities with regional United States Postal Service officials and with military and commercial transportation managers, and to provide detailed and timely instructions to the United States Postal Service officials for processing, routing, and dispatching Military Postal Service mail.

Mail is a morale booster for troops fighting overseas and for their families at home. More than 65 million pounds of letters and parcels were delivered to troops serving in Operation Iraqi Freedom in 2003 and problems with prompt and reliable mail delivery surfaced early in the conflict. Congress and the White House forwarded more than 300 inquiries about mail delivery problems to military postal officials.

The Government Accountability Office was directed to review mail delivery to troops stationed in the Middle East. In this report, the Government Accountability Office assessed the timeliness of mail delivery to and from troops in Operation Iraqi Freedom, how mail delivery issues and problems during this operation compared with those experienced during Operations Desert Shield/ Storm in 1991, and efforts to identify actions to resolve problems in establishing mail operations for future contingencies.

The timeliness of mail delivery to troops serving in Operation Iraqi Freedom cannot be accurately assessed because the Department of Defense does not have a reliable, accurate system in place to measure timeliness. In general, the Department of Defense's transit time and test letter data show that mail delivery fell within the current wartime standard of 12 to 18 days. However, the methodology used to calculate transit time significantly understated actual delivery times.

Despite differences in operational theaters and efforts by the Department of Defense postal planners to incorporate Operations Desert Shield/ Storm experiences into planning for Operation Iraqi Freedom, postal operations faced many of the same problems;

difficulty with conducting joint-service mail operations, postal personnel who were inadequately trained and initially scarce owing to late deployments, and inadequate postal facilities, equipment, and transportation. The operations plan created for joint-service mail delivery contained certain assumptions key to its success but led to unforeseen consequence or did not occur. Also, plans for a Joint Postal Center were not fully put in place. One lesson learned from 1991 was carried out with success during Operation Iraqi Freedom; mail was transported oversees by dedicated contracted airlifts rather than by military.

The Department of Defense has not officially tasked any entity to resolve the long-standing postal problems experienced during contingency operations. Moreover, the Military Postal Service Agency does not have the authority to ensure that these problems are addressed jointly. The agency and the military services, however, have taken some steps toward tackling these issues.

The Government Accountability Office has recommended that the Secretary of Defense implement a new system to accurately track, calculate, and report postal transit times and consolidate lessons learned from Operation Iraqi Freedom and develop and implement a specific course of action to resolve them. The Department of Defense, in its formal review of this matter, fully concurs with the Government Accountability Office's recommendations and has begun taking steps to implement them.

The Military Postal Service Agency is The Department of Defense's single manager for military postal functions. This joint service agency is organizationally located under the Army Adjutant General and depends on the Army for funding and staffing.

The movement of mail from the United States to troops in the Iraqi theater follows several complex logistical steps. Letters and parcels with military addresses destined to Iraq, Kuwait, and Bahrain are sent to one of four International Mail Gateways in New York, San

Francisco, Chicago and Miami for processing. The United States Postal Service delivers letters to the International Service Center at John F. Kennedy Airport, in New York. Parcels are delivered to the Postal Service's International and Bulk Mail Center in New Jersey. After the letters and parcels are sorted, they are then packaged, placed into containers, and then transferred to Newark International Airport in New Jersey where they are loaded onto airplanes for transport to the Iraqi theater.

Unlike during Operations Desert Shield/Storm, where military planes operated by the Military Airlift Command transported much of the mail, a dedicated contractor aircraft carries mail during Operation Iraqi Freedom. During the next stage of mail movement, the mail planes fly to aerial mail terminals co-located at the international airports in Kuwait and Bahrain. Once landed, local airport ground handlers offload the mail containers from the planes and take them to an Air Force Mail Control Activity located at the airport, where the mail is staged for ground transportation. In Bahrain, mail for service members stationed in Iraq is processed at the United States Air Force Mail Control Activity. Mail for service members located in Bahrain or aboard ships is processed at the United States Fleet Mail Center. For troops stationed in Iraq, mail is transferred onto a contracted cargo plane and flown directly into Iraq.

The Joint Military Mail Terminal, which handles the bulk of letters and parcels entering the Iraqi theater, sorts the mail and arranges for its transportation either by land or by air to the various regions occupied by the United States troops. Mail must be delivered to the unit level, designed by ZIP codes provided by the Military Postal Service Agency, before it can be distributed to individual service members.

The biggest hindrance to processing mail is a lack of sufficient workspace. Without taking action to resolve the identified issues in planning, building, and operating a joint postal system, mail delivery will continue to suffer in future contingency operations as witnessed by the repetition of delayed mail delivery from one Gulf war to the next. Emphasis needs to be placed on establishing joint postal responsibilities and the subsequent execution of those duties. Past experience has shown that postal operations have not received command attention or been designated a priority. Establishing the needs for postal operations early in the process and dedicating the appropriate resources is crucial for providing the timely and efficient delivery of mail. While our work focused only on Operation Iraqi Freedom, we believe many of these same lessons apply to other combatant commands and theaters of operation as well. Without clear and accurate data to measure the timeliness of mail to United States troops overseas during contingency operations, no meaningful assessment can be made on the quality of mail service. Therefore, it was recommended that the Secretary of Defense work with the Army Adjutant General to improve the quality of transit time data for postal operations by implementing a system that will accurately track, calculate, and report postal transit times. In the absence of a clear plan for resolving recurring postal problems during contingency operations, it was recommended that the Under Secretary of Defense (Acquisition, Technology, and Logistics) designate, direct, and authorize an appropriate Department of Defense agency, unit, or command to determine what long-standing postal issues need to be resolved, and to develop a specific course of action and timetable for their resolution, including appropriate follow-up to ensure that the problems have been fixed. Specifically, these actions should address the issues highlighted today such as strengthening the joint postal planning function and specify a body to ensure the implementation of postal operations in theater, deploy properly trained and equipped postal troops into theater prior to the mail build-up, and dedicate adequate postal facilities, heavy equipment, and transportation assets for postal operations. An important part of addressing these long-standing problems is to share the results of these lessons learned from Operation Iraqi Freedom with all.

The Military Postal Service has come a long way since the beginning of the war. Since then mail delivery has been more smoothly and has been delivered in a more efficient timely manner. As it stands, it takes about eight to ten days to deliver mail from theater to the United States. Mail from the United States to Iraq generally takes about the same to reach the Army Post Office in theater. It is then up to the individual units to disperse the mail to their Soldiers in a timely manner. These numbers do fluctuate some during the holidays and busier time of the year.

The Military Postal Service

An Army of Air Force
Written by: LT Trent McGee, LT, USAF, 732 ESS (Postal)
International Zone, Baghdad, Iraq

"Yes" was the only word out of my mouth when my commander asked if I wanted to deploy. I didn't know where I was going or what I would be doing, but I knew that I wanted and need to be serving overseas. With the recent cuts in the Air Force and 'perceived' views of certain lieutenants at my base I was fully aware of the possibility for my letter to step down. There was no way I was going to leave without a deployment, so when I volunteered, I was happy that something came down for me. It turned out to be an Army in-lieu-of tasking and I had roughly four days to prepare for deployment.

Two and a half months of Army training dragged by, WHOOA, before I was finally dropped off in Baghdad. I firmly believe the Army requires so much pre-deployment training because they know after a month or two of training all you want to do is deploy so you can get away from it. Sure, there are other benefits, like preparing you to kill, rather than be killed, but with all that aside, I was ready to be done with training and start working. I was tasked to be a Postal Officer at Union III Post Office in the International Zone, Baghdad, Iraq. I was new to the Postal world but I was optimistic and excited about my first deployment. Not only was I deployed to the heart of Iraq but I was deployed to fill an Army slot, which always intrigues people when they hear it. It also intrigues me. Why would the Air Force continuously remove manning and send more and more troops to support the Army? If we are removing people from the Air Force altogether and sending people to fill Army slots, who is going to perform the Air Force mission? Then I hear the cliché "Do more with less" and remember that technology is going to make all of our jobs easier. You know like having e-mail and cell phones has made it easier for you to work no matter where or what your vacation plans include. Anyways, back to the topic, the first 30 days were an eye opener to me and my Airmen. I had 16 personnel under my command and five of them were in the postal career field. While we all had some training, as I mentioned earlier, my five core postal Airmen were working overtime to ensure that the rest of the Airmen including myself knew what was going on. I had an Army and Air Force chain-of-command located in two different FOBs other than my own, communication was non-existent because we arrived at the same time the Army units were transferring authority, and we were rapidly approaching Christmas, the busiest time of year for Postal. All in all it was exactly what I was looking forward to in a deployment, a challenge.

Our initial experiences with the Army were as one might expect, rough. As members of the military we are all used to the inter-service colloquialisms, and this deployment did not disappoint. Minus an officer or two and the occasional Office of Special Investigations member we were the only Air Force members on FOB Union. The looks and random "Aim High" yelled out at the DFAC were quickly responded to with a "WHOOA" or "Wouldn't want to hit you" during the first few weeks, but it quickly died down once people realized we were there to provide their mail. I do not know if it is because we handled mail and it was Christmas time or whether the Army just started taking a liking to us, but things were much better after people started receiving their packages. I can say that it was never construed as anything other than joking around and I was grateful for that as the Postal Officer. The last thing I needed were issues between my Airmen and the Soldiers. There were enough issues being a commander of sixteen people, with sixteen different personalities that came together for this deployment from many different locations. I had Airmen from all over the world, Germany, Norway, England, California, Hawaii, and Delaware just to name a few. Granted we were at the same training but they were not under my command at the time and that is a story that is better left untold.

Before I left for the deployment I heard many statements along the lines of "Oh you are going to the IZ, that is a sweet assignment, you will have your own trailer and wireless internet and there is a pool, etc.....". Well, they were wrong. They were under the impression that I would be staying at the Embassy compound (where the majority of Air Force can be found). FOB Union III, while only a mile or so from this oasis, did not have quite the same luxuries. It is in every aspect a fully functioning Army FOB. If not having something won't kill you, then you probably don't need it. Many of the complaints I heard were about paying for internet service to be in your room, nothing to do with having to travel to the shower and bathroom, and not having enough space. I loved it. Deployments aren't supposed to be easy. Granted the complaints that I listed will just get a confused "are you actually saying this to me" from an Army E-7 or above, it is quid-pro-quo in the Air Force, so I fielded the complaints and let everyone settle into their new environment. I knew that my Airmen weren't really upset by these trivial things, they just needed time to adjust to the deployed lifestyle.

The deployment offered many firsts for me as a leader. I had an Airmen break her arm and needed surgery in the States. I processed paperwork for my first Article 15 and juggled with two chains of command depending on the situation. Throughout these experiences I noted several key differences between the Air Force and Army. Before I go into the difference, I should reiterate that the Air Force and Army have inherently different mission, so in no way should the two branches be the same. These observances are just my opinion, as someone experiencing both branches at the same time on a day to day basis, and for the purpose of this article alone. The first thing I noticed is that while the Air Force teaches its Airmen to respect and follow the orders of their superiors it also advocates voicing their ideas. The Army, on the other hand, is very straight forward. When something needs to be done, the order is given and it is done the way the SNCO or Officer said to do it. There are good and bad for each method but

I am not here to judge them, just state what I observed. The second major difference that I noted deals with punishment. From what I have observed, in my short time as an officer, the Air Force requires paperwork for any type of punishment. In order to justify something as small as push-ups, it must be documented. The means trivial things go unpunished because it is not worth giving someone a Letter of Counseling. I saw several Soldiers pushing their way out of trouble though. Even small measures like that can be a deterrent. The last thing I noticed that I am going to speak about is regarding briefings and reports. The Air Force has the usual safety briefings, reports and online training, but the Army requires a report or spreadsheet for almost everything. I understand the importance of accountability of personnel and equipment, and I am sure the Army has many reasons why it conducts business the way it does. It is possible that it seemed like more to me because I was reporting up two chains of command. As I stated earlier, I am not criticizing or praising either side in this article I just found these characteristics of the branches had the most contrast.

In all, working for the Army was a very pleasant experience. We were recognized as Airmen but respected as Soldiers. The leadership was there for any questions that I might have had and supported us every step of the way. I was not worried about being pawned off if there was an issue, which can happen when there are two sides working together. The Army accepted us into their unit and made us feel right at home. Currently there are thousands of Air Force personnel working ILO Army taskings and I hope that they have the same experience I had. I have been in the Air Force for 3 years in June. I will always be able to say that the first medal I received as a member of the Armed Forces was from the Army. Working for the Army has given me and my Airmen a chance to see how our brothers and sisters in the Armed Forces perform their mission. I can honestly say that I am proud of all the Airmen that served with me and I am grateful to the Army and the Soldiers that we served with for making this experience what it was. I will always remember this deployment since it is most likely my last. I could not have asked for better people to work with and I appreciate everything the Army did for us as a unit. I am grateful for my experience and hope everyone makes it home safely.

The Military Postal Service

Postal Challenges and Opportunities
Written by: 1LT Allison Morse

While deployed to Iraq: What can be mailed to the United States for free? What items cannot be mailed at all? What are the weight and dimension limitations for an outgoing package? These questions, and many more, are what my Soldiers and I are asked on a daily basis. The answer to these question are all aspects of our combat mission in Baghdad, Iraq. My current duty position is as a Postal Platoon Leader for the 3/129th Adjutant General Company (Postal), deployed to Iraq in support of Operation Iraqi Freedom 06-08. During deployment the challenges of overseeing a post office and the various stages of Army transformation have helped shape my leadership style and helped prepare me for future assignments.

As a Postal Platoon Leader I am in charge of several postal facilities and operations that service approximately 17,000 Soldiers, DoD Contractors, and Civilian Contractors on the Victory Base Complex in Baghdad, Iraq. Our main facility has a postal finance and postal operations section. Postal finance is further broken into the inspection section, the finance section where mail postage is paid for, and the claim inquiry section. The postal operations sections includes a myriad of tasks. It is here that mail is sorted to the over 180 units on and surrounding our camp, accountable mail is separated from regular mail, mail is redirected to Soldiers who have moved or redeployed, and mail clerks undergo the certification class. I am also responsible for a satellite post office at a smaller, nearby camp. This post office only has a finance section which performs the same tasks as the main post office's finance section. In addition, we conduct two postal mobile missions. By adding these missions we are able to take the post office to the customer who would otherwise be unable to make it to the post office during business hours.

I learned quickly upon arrival to the Camp Victory Army Post Office that in order to be successful I would have to know the regulations inside out, and where to locate the correct answers. The same, I am sure, is true for any other AG position. The most common challenge occurs when a customer does not understand the Army Post Office regulations. This may include what constitutes free mail, why certain items cannot be mailed, or questions on why mail may take longer than normal to go from Iraq to the United States. When these questions are asked, it is not enough that I know the answer. I must also be able to show the customer the answer in the regulation. This does not mean that I know the exact page number, but I pride myself in knowing whether to look in DoD 4525.6-M or the Consolidated Air-Massing and Labeling Scheme (CAMALS). Another aspect of this is ensuring that my Soldiers know the correct answers. And where to reference the information. My NCOs and I accomplish this mission through classes and supervision to ensure that they not only know the regulations and new policies, but are also following those regulations. Another reason that it is important for us to know the policies is because we are not isolated from the post office and even though there may be distance between us. Our customers utilize the post offices at the various camps, and expect the same standards from each post office. So it is important that the leadership knows the policies while supervising the Soldiers, the Soldiers know the regulations, and that we keep in contact with our fellow postal officers to ensure we are following the same Standard Operating Procedures.

Another challenge we face is that there always seems to be limited room for the massive amounts of mail we encounter. This is especially true during the holidays, redeployments and with the recent surge of troops. This is when we must step away from the regulations, and use creativity to become the masters of our space and resources. We utilize our outdoor space for additional inspection space, and section off space for additional mail trucks to park. By rearranging the units in various sorting bins we are able to maximize our indoor space. It is this out of the box thinking that allows us to maximize our limited space.

An additional issue is personnel. This is a valued resource that I am sure all branches would like more of. For the three years prior to the OIF 06-08 rotation this post office has always been manned by two platoons. However, on this deployment there is only one platoon, and in fact even more mail to sort. This is an issue that several other post offices encountered as well. Mail sorting was especially difficult during the Christmas season, but with the aid of augmentees tasked to the post office the mission was a little easier. Another critical piece that led to the post office's success, and in an enduring contribution, is the professionalism, discipline, and loyalty of well-trained Soldiers. Because of these characteristics continuous supervision is unnecessary, and this allows the NCOs and me to focus on improving the post office and the quality of life for our Soldiers.

At this time the Camp Victory Army Post Office appears to be a work in progress in several aspects. There is the continual rotation of units, the upgrades to the building that was not designed to last this long, and the addition of civilian contractors. A change in units means that sort bins must be rearranged, Soldiers must learn the new bin locations, and our technical inspector must teach more mail clerk classes. We have taken measures to increase security and safety measure by emplacing tall walls, making ramps around the facility, and creating a safer, more even area where the trucks drop off mail. Our post office is also in the process of undergoing a vast transformation. Our current goal is to transfer the entire military staff postal operations over to civilian contractors. In the end state there would still be a military OIC, NCOIC and Custodian of Postal Effects, but all other positons would be held by civilian contractors. This would allow our Soldiers to concentrate on other functions in the Army. The step that we are currently working on is transferring the postal operations section to civilian contractors at their facility. This would increase mail service in several ways. Some of the benefits are that the contractors have a larger space, an updated facility and equipment in better condition to conduct the mission. Also, they would have more personnel to dedicate to the mail service. All of this would lead to faster service for the Warfighter. This part of the transformation should run rather smoothly. Civilian contractors already sort the mail to the various APOs. This step basically requires that they go one step further and sort to the units at the APOs. Once civilian contractors take over the operations sections they can begin to focus on transferring the finance section. Another future possibility may even be moving the finance section to an updated facility.

I believe that my postal experiences have had a significant impact on my leadership style, confidence, and helped groom me for my future positions whether they be in the Army or civilian sector. Being a postal platoon leader is the only platoon leader position in the AG branch. I feel fortunate that I was given this opportunity. Throughout college I was exposed to various leadership styles. In those classes I remember thinking I would demonstrate a participative versus directive style. I also thought I would be a great leader because of my compassion and my ability to relate with other people. This platoon leader position allowed me to exercise several different styles, and helped me realize which style was truly best for me. I am now more of a directive leader. I found out the hard way that young Soldiers, and even some NCOs, find compassion and a participative style to be a sign of weakness. I learned very quickly that if I leave my decision up for discussion then Soldiers will always counter with why they do not like the idea. When dealing with some Soldiers even using reason will not work, and they simply need direction. I no longer feel compelled to give an explanation behind my decisions or gather input from everyone involved. I realize that I don't know everything, and should take advice from my NCOs, but I now understand the decision is ultimately mine. I must make decisions based on their advice, my knowledge, and my experiences, and as long as my decisions is not immoral, illegal or illicit I have the final say. Now I start every relationship with one of my Soldiers as directive. I tell them exactly what I want them to do, and how I want it done. If the Soldier can then demonstrate that they have an understanding of the standards that I expect, then I give them room to decide how to fulfill the mission. However, I continue to supervise.

In conclusion, running a post office is not a simple task, but one that I have found critical to my growth as a leader. Knowing the regulations, thinking creatively, maximizing limited resources, and always looking for ways to improve service to the customer was a challenging process that allowed my Soldiers and I to have a significant contribution to the morale of the Warfighters serving in support of Operation Iraqi Freedom 06-08. This combat mission was not always easy, and I attribute the Camp Victory Army Post Office's success to the professional, discipline, knowledge and loyalty of the Soldiers and NCOs of the 3/129th AG Company (Postal).

The Military Postal Service

Postal Operations and Mail Flow in the Iraqi Theater
Written by: 2LT Steven Druckenmiller

One could make the case that in terms of Soldier morale and well-being, mail would fall fourth in the order of precedence, immediately after water, food and shelter. To the American Soldier, mail is a sign that someone who loves them took the time to think about them during this stressful period. Given the significant impact mail has on mission effectiveness, it is incumbent upon those responsible for mail flow in theater to do the best job they can. That is where the 15th Personnel Services Battalion (PSB) enters the picture.

The 15th PSB, or the "Broncos!" from Fort Hood, TX, (Commanded by LTC Angela M. Odom) is responsible for approximately two-thirds of the personnel services in the Iraqi theater; running essential personnel services for 13 out of 20 BCTs currently forward deployed to this area of operations. The most highly-visible mission the 15th PSB runs is the Postal mission. The bottom line up front is that the 15th PSB and its subordinate postal companies are responsible for the mail for over 96,000 American Soldiers, Sailors, Airmen and Marines. With this humongous customer base, the number of locations where the 15th PSB runs mail operations and the importance of the mail, the mission is a huge responsibility.

How exactly does mail flow work in Iraq? Let us start at the beginning. When a Family member mails a letter, which travels to the Joint Military Postal Activity in New York (located at John F. Kennedy International Airport; herein abbreviated as JMPA-ATL). This is commonly referred to as the "gateway" for the mail (there are three in the United States: JFK, San Francisco, called JMPA-PAC, and one that services U.S. Soldiers stationed in SOUTCOM, located in Miami). JMPA sorts the mail for Bahrain. Bahrain workers sort the mail according to the Joint Military Mail Terminal (JMMT) it will be traveling. There are three JMMTs in the theater: there is the JMMT-K (Kuwait), the JMMT-N (North – located in LSA Anaconda, commonly called Balad) and the JMMT-S (South – located near Baghdad International Airport).

So, how does the mail get from Bahrain to the various JMMTs? After traveling from JMPA-ATL to Bahrain via various air carriers, the mail is sorted and palletized. International mail and package carrier SHL (Dalsey, Hillblom and Lynn, a subsidiary of Deutsche Post, the German post office) flies the mail into the three separate JMMTs.

Soldiers at JMMT-South (Baghdad International Airport) are starting a brand-new operation and concept. At the JMMT, Postal Operations Soldiers from the 175th AG CO (Postal) (DS) sort the mail for specific FOBs (Delineated by APO ZIP). Prior to the beginning of the "JMMT initiative," trucks would convoy the mail to the separate APOs in the BIAP area. From there, Unit Mail Clerks would travel to their specific APO to receive the mail for their Soldiers. Now, in the BIAP area, all Postal Operations and Mail Pick-Up have been consolidated into one location: JMMT-S. This initiative helps to maximize the efficiency of the individual Postal Operations Soldier. Ultimately, the Postal Directorate is looking to contract this initiative in order to free up highly trained Postal Soldiers for more essential missions.

In terms of Postal capabilities, the 15th PSB already goes above and beyond its doctrinal abilities. The customer base, as stated earlier, is approximately 96,000 U.S. Forces and contractors. Doctrinally speaking, this would support having 17 Postal Platoons (One 18-man platoon for each 6,000 customers) in this theater of operations. The 15th PSB, with only 13 Postal Platoons, not only fully completes its doctrinal Postal mission and the requirements; it frequently goes "the extra mile" to totally support the Warfighter in theater. These 13 Postal platoons receive Command and Control from two exceptional Postal Companies; the 175th AG CO (Postal) (DS), commanded by

CPT Dale Jones, is an active duty Postal Company from Fort Lewis, Washington. CPT Kenneth Fillmore commands the 834th AG CO (Postal) (DS), a USAR Postal company from Miami, FL, under the 81st RRC. Although the command teams of these companies are made of Soldiers from both the Active and Reserve Companies.

The 175th AG CO (Postal) (DS) is headquartered in Baghdad. Its primary responsibility is mail operations in Baghdad and the surrounding area. This includes the camps surrounding BIAP, the JMMT-S and various Camps and FOBs within rotary-wing distance of Baghdad.

The 175th AG CO (Postal) (DS) commands eight platoons: as stated earlier, about an even distribution of Active and Reserve Component platoons. 175th AG CO (Postal) (DS) is also notable for having an Air Force Postal Platoon under its command and control. In addition to the great responsibility of servicing the most highly visible area in the world right now, the 175th AG CO (Postal) (DS) also has the distinction of handling some of the largest amounts of mail in the theater. On average, the 175th AG CO (Postal) (DS) processes anywhere from 500,000 to 1,000,000 pounds of mail per week. The 175th AG CO (Postal) (DS) has been in theater since October 2006, and has already processed over 16,000,000 pounds of mail and generated almost six million dollars in revenue from postage sales and money order services. With the Baghdad security plan coming into full force, the 175th AG CO's average of 4,500 customers per week is with expectation to increase significantly.

The mail flow for this unit's sector is a veritable frenzy of activity as mail is pushed from the JMMT-S to the various FOBs and Camps near Baghdad, typically via KBR or logistical convoys. Many of the APOs in the area run 24 –hour, 7-day a week operations, due to the variability in scheduling of these Convoy Logistical Patrols (or CLPs). Major operations areas include not only the JMMT-S but Camp Taji as well. When the mail reaches its particular APO, the mail is sorted again, this time by unit, at the APO, and that is when Unit Mail

Clerks receive their mail during the day and distribute it to individual Soldiers.

Postal Operations run slight differently in the southern portion of Iraq. Here, we find the 834th AG CO (Postal) (DS) and its commander, CPT Fillmore, running a very busy but geographically dispersed operation. The 834th AG CO (Postal) (DS) runs Postal Operations from the west of Iraq (Al Asad and Ramadi) to as far south as the border of Kuwait (Camp Bucca). The 834th headquarters is in LSA Adder (or Camp Tallil). Although the operation is primarily in the less-populated south, the diversity of geography in the operation means theater the 834th AG CO (Postal) (DS) put up numbers that rival its brother Postal Company. The 834th AG CO (Postal) (DS) has been on the ground since September 2006, and in that time they average anywhere from 300,000 – 400,000 pounds of mail processed per week. In the time that they have been in theater, they have generated over $900,000 by selling postage and processing money orders. They have serviced almost 30,000 customers and processed approximately six million pounds of mail (incoming and outgoing). The mail for the 834th and its area of operations comes on combat logistic patrols (CLPs) from the JMMT-K, with Tallil being the primary processing center for many of the FOBs and Camps that the AG Company services.

In order to provide services to areas that cannot support (due to customer base or lack of facilities) a permanent postal presence, the two Postal companies in the 15th PSB's Area of Operations perform what are called "Postal Rodeos." Rodeos are movements of a team of Postal Finance Soldiers from a Camp or FOB with an Army Post Office to an outlying area that needs Postal Service. Occurring anywhere from twice a week to once every month, these Rodeos ensure that the 15th PSB is servicing all of the Soldiers in the Battalion's footprint, giving many opportunities to the Warfighter to mail packages home to their loved ones.

In summary, mail operations require sustained and exceptional dedication, hardworking Soldiers and thousands of people to make it effective. The mail is moved by airplanes, helicopters, tractor-trailers, 5-ton trucks, forklifts and, of course, the hands of Soldiers across Iraq. Army leaders' input and effort, from the highest ranks in the Pentagon to the section leader in the APO, is vital and necessary to ensure that their morale-boosting and necessary wartime function are executed safely, efficiently and, of course, "First Class!"

Casualty Operations in Iraq

Written by CPT Richard M. Strong, Charlie Detachment, 15th PSB, and 2007 AGCRA Officer of the Year

Casualty operations have changed a great deal since the beginning of Operation Iraqi Freedom (OIF) in 2003 to the current rotation today. As the Commander for Charlie Detachment, 15th Personnel Services Battalion, my unit is responsible for several Casualty Liaison Team (CLT) missions in support of OIF 06-08. We deployed in October 2006 and originally had ten Soldiers immediately move into the Multi-National Division – Baghdad (MND-B) headquarters to perform as a CLT. Charlie Detachment also had a four- Soldier CLT move to one of the busiest Combat Support Hospitals in theater, the 28th Combat Support Hospital (CSH) in the International Zone. When the 3rd Infantry Division deployed and activated Multi-National Division – Center (MND-C), seven PSB Soldiers moved into their headquarters building and performed as the Division CLT. However, CLTs out of Human Resource Companies arriving into theater in support of OIF 07-9 will be replacing our CLTs in the divisions, the combat support hospitals, as well as in Multi-National Corps-Iraq(MNC-I), CJ1.

There were several ways a casualty was reported in the past. The battalion Tactical Operations Center (TOC) would send up a Significant Acts (SIGACTs) report to the brigade TOC via Command Post of the Future (CPOF), to division level, and eventually to corps level. The battalion or brigade S1 would report any causalities using the basic data (11-line initial casualty report). The 11-line initial casualty report contained the immediate and essential information required to start the casualty notification process. It included information such as the name, rank, unit, type of casualty, circumstances/ remarks, etc. The brigade level S1 would forward the report to division level G1, as well as a CLT sourced by a personnel services battalion. Division G1 would take that information and track the casualty to keep the division commander and command group informed at all times. The Deployed Casualty Information Processing System (DCIPS) was completed by the Casualty Cell manned by Soldiers from the PSB and forwarded up to the casualty section at Multi-National Corps-Iraq (MNC-I). MNC-I reviewed the DCIPS report and would forward it to Coalition Forces Land Component Command (CFLCC) and the Casualty Assistance Center (CAC) in Kuwait. The CAC in Kuwait transmitted the DCIPS to Casualty Memorial Affairs Operations Center (CMAOC) and eventually back to the CAC at home station, which started the casualty notification process.

The Personnel Services Battalion CLT was originally collocated within the Division Headquarters to allow synchronization between the CLT, the Division G1, and the Division Surgeon Section (DSS) in order to provide the CLT with additional sources to receive information. The Division G1 provided the SIGACTS report and currently updates on incidents occurring in their Area of Responsibility (AOR). The DSS were able to provide updates on the progress of causalities as well as verify the casualty information by cross-referencing patient information submitted by the medical facilities. The CLT located at the 28th CSH also contacted the Division CLT when they received patients in the CSH. They also assisted the Division Surgeon Section with updated patient status information.

The Personnel Services Battalion CLT located at the 28th CSH was responsible for accomplishing several tasks. Their main task was to complete spot reports as casualties were admitted into the CSH and send those reports to the CLT at Division to complete the DCIPS report. They also made rounds around the different wards of the hospital and updated the patients' statuses. Eventually, the CLT at 28th CSH will complete DCIPs reports and send them to Division and MNC-I.

The Personnel Services Battalion is one of the last of its kind left in the United States Army. They have been phased out in accordance with the Personnel Services Delivery Redesign (PSDR) concept. PSDR is the Army Human Resource (HR) community's functional realignment to synchronize with the Army's transformation. It eliminates support layers and minimizes the support unit footprint in the battle space. PSDR embeds critical military personnel services functions in the brigade S-1 section and empowers commanders, in Garrison and in Theater, to provide Human Resource support directly to their Soldiers. The brigade combat teams (BCTs) will be able to complete tasks the Personnel Services Battalions use to complete. One of those tasks is to create DCIPS reports. When the BCT S1 completes the initial DCIPS report, they will send it to the CLT at division level for accuracy screening, who will then send it up to MNC-I, CJ1 Casualty Section.

The transition of brigades submitting the 11-line report to the division CLTs to creating and submitting the DCIPS report is not an immediate, one-step transition. As soon as 3rd Infantry Division was scheduled to deploy as a second division headquarters in the Baghdad area, the units designated to be task organized under the new Multi-National Division-Center needed to be trained on DCIPS reporting procedure. Charlie Detachment, 15th Personnel Services Battalion assembled a training team that travel down to those brigade headquarters to instruct their S1 sections on DCIPS reporting. The training lasted approximately three days and consisted of a PowerPoint slideshow presentation, handouts, and practical exercises. The PowerPoint slides and handouts included detailed guidance and instructions on DCIPS reporting procedures, possible unfamiliar vocabulary and actual experiences. The instructors also assisted with setting up the available brigade S1 area. The Day 1 training focused on Wounded-in-Action (WIA) DCIPS reporting and concluded with practical exercises and one-on-one training. Day 2 training focused more on deceased DCIPS reporting because they are more in depth and require more information. Day 2 also concluded with practical exercises and one-on-one training. Day 3 (and anytime the unit received an actual casualty throughout the training), the trainers supervised and assisted the brigade S1 section in creating a real DCIPS report for submission to the Division CLT. After the training was complete, the brigade S1 sections would continue to send up the 11-line report and a DCIPS report to the division CLT for about one week. This was to ensure the brigade S1 was submitting the DCIPS repots correctly. The division CLT would review and verify the report and then send it forward to MNC-I, CJ1 Casualty Section.

The desired-end-state for the current rotation is to closely mirror incoming replacement CLTs from Human Resource Companies. The replacement CLTs will consist of five Soldiers, one of which will be a Noncommissioned Officer. The Human Resource Company CLTs will be positioned at each Level III medical facility (28th CSH for example), each Division G1 (Multi-National Division – Baghdad and Multi National Division – Center for example), and at Corps CJ1 Casualty Section. Therefore, the CLT strength needs to eventually be scaled down to match the five Soldier strength that will be replacing the currently CLTs. The existing embedded casualty cell co-located with MND-B Headquarters has nine Soldiers and Noncommissioned Officers because they are still tracking the 11-line report from brigades and creating the DCIPS reports at their level. Planning and scheduling is currently taking place to get out to the brigade headquarters locations and train them on DCIPs reporting. Once MND-B subordinate brigades are submitting DCIPS reports, the Casualty Cell troop to task will decrease significantly. Instead of creating the DCIPS, they simply verify the report and check it for errors.

The new doctrinal use of CLTs should speed up the current casualty reporting process without compromising accuracy. When a casualty occurs in a unit, the battalion Tactical Operations Center (TOC) will send the Significant Acts report to brigade via Command Post of the Future (CPOF) through Division to the Corps. The brigade S1 will take the basic data (11-line initial casualty report) and compose a DCIPS report. That DCIPS report will be submitted to the CLT co-located at Division Headquarters, who verifies the information and further submits it to MNC-I CJ1 CLT. By this point, the actual casualty has probably reached the CSH where another DCIPS report is generated and submitted to the Division CLT and carbon copied to Corps. It is possible that the DCIPS report from the CSH arrives at the Division CLT before the DCIPS report from the unit. Multi-National Corps-Iraq is the clearinghouse for all casualty reports before they leave theater. The multiple DCIPS reports on one casualty are synthesized and one DCIPS report is sent to Kuwait CAC which transmits to Casualty Memorial Affairs Operations Center (CMAOC).

While every Soldier in Charlie Detachment, 15th Personnel Services Battalion has had to remain flexible in their current duty position, one thing has always remained the same: Casualty Operations is one of, if not the most important mission in theater. It deals with the most sensitive information and has the highest visibility of any other subject in Iraq. Any casualty report leaving theater has to be absolutely and unequivocally error-free because our Soldiers and their families deserve timely and accurate notification. Erroneously notifying a Soldiers' Next of Kin that their son or daughter has been wounded or killed in action is unacceptable. Having a Family member find out that their son or daughter has been injured or killed on CNN or Fox News instead of the proper casualty notification channels is also unacceptable. In the end, transitioning into the new five Soldier Casualty Liaison Team model with initial DCIPS reports created at brigade level and at the combat support hospitals giving MNC-I the most accurate data in the shortest amount of time.

TASK FORCE BRONCO'S PICTURE

Concrete bunkers were used for force protection on Victory Base Complex, Camp Liberty, Iraq, Pad 11

15th PSB's unit sign mounted in front of the battalion headquarters on Victory Base Complex, Camp Liberty, Iraq, Pad 11

Task Force Bronco Soldiers transporting the mail to remote Forward Operating Bases

15th Personnel Services Battalion's Headquarters from October 2006 - January 2008; Victory Base Complex, Camp Liberty, Iraq, Pad 11

The Command Team for Task Force Bronco during Operation Iraqi Freedom (OIF) during October 2006 through January 2008, COL (R) Angela M. Odom (Battalion Commander) and CSM (R) Timothy P. Livengood (Battalion Command Sergeant Major)

Unit sign outside of the 15th Personnel Services Battalion's headquarters located on Fort Hood, TX. The Army deactivated the unit in 2008.

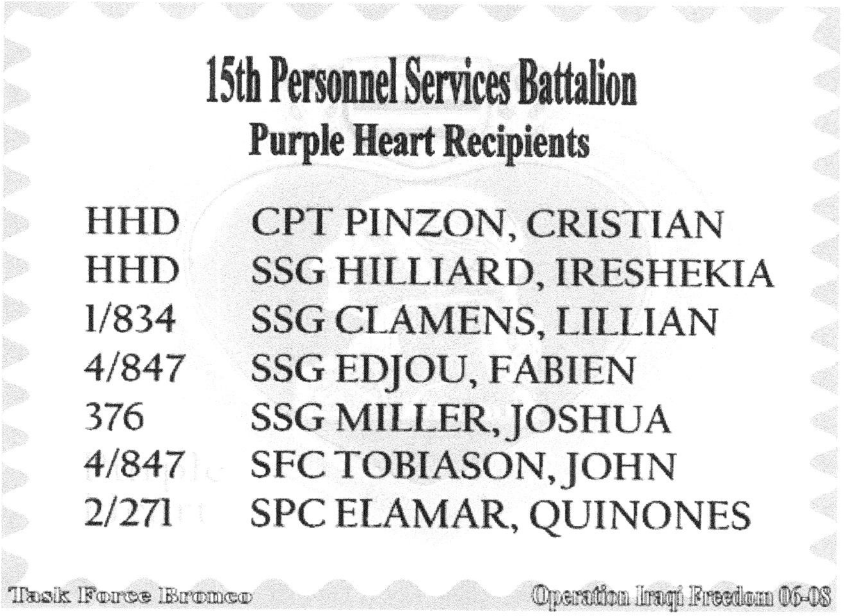
15th Personnel Services Battalion
Purple Heart Recipients
HHD CPT PINZON, CRISTIAN
HHD SSG HILLIARD, IRESHEKIA
1/834 SSG CLAMENS, LILLIAN
4/847 SSG EDJOU, FABIEN
376 SSG MILLER, JOSHUA
4/847 SFC TOBIASON, JOHN
2/271 SPC ELAMAR, QUINONES
Task Force Bronco
Operation Iraqi Freedom 06-08

TASK FORCE BRONCO'S PURPLE HEART RECIPIENTS

The Purple Heart was established by General George Washington at Newburgh, New York, on 7 August 1782, during the Revolutionary War. It was reestablished by the President of the United States per War Department General Orders in 1932. The Purple Heart is awarded in the name of the President of the United States and per 10 United States Code 1131, effective 19 May 1998, is limited to members of the Armed Forces of the United States who, while serving under component authority in any capacity with one of the U.S. Armed Services after 5 April 1917, has been wounded or killed, or who has died or may hereafter die after being wounded.

In the journal of the Adjutant General's Corps Regimental Association's *1775* Special Edition, titled, *Heroes Among Us* it is noted that a total of 347 Adjutant General Corp's Soldiers have earned the Purple Heart medal since the Vietnam era.

Since the tragedy of 9/11, 33 Adjutant General Corps' Soldiers have made the ultimate sacrifice to our Nation and Army was either killed in action or was a fatality supporting Operation Iraqi Freedom and Operation Enduring Freedom. Unfortunately, *two* of them were members of Task Force Bronco in 2007.

During Operation Iraqi Freedom 06-08, seven members of Task Force Bronco were award the Purple Heart.

Task Force Bronco's Purple Heart Recipients Are:

*SSG Clamens, Lillian L. 1st Platoon,
834th Adjutant General Company (Postal) (Direct Support)

SSG Edjou, Fabien
Detachment 4, 847th Personnel Services Battalion

SPC Elamar, Quinones
2nd Platoon, 271st Adjutant General Detachment

SFC Hilliard, Ireshekia
Headquarters and Headquarters Detachment
15th Personnel Services Battalion

SSG Miller, Joshua
Detachment 5, 376th Personnel Services Battalion

CPT Pinzon, Cristian
Headquarters and Headquarters Detachment
15th Personnel Services Battalion

*SFC Tobiason, John J.
Detachment 4, 847th Personnel Services Battalion

*deceased

NOTE: This listing is based on internal record keeping.

TASK FORCE BRONCO'S COMBAT ACTION BADGE RECIPIENTS

The Combat Action Badge (CAB) is awarded to Soldiers who are personally present and actively engaging or being engaged by the enemy, and performing satisfactorily in accordance with the prescribed rules of engagement.

During Operation Iraqi Freedom 06-08, thirty-nine members of Task Force Bronco earned Combat Action Badges.

Task Force Bronco's Combat Action Badge Recipients Are:

Headquarters and Headquarters Detachment, 15th Personnel Services Battalion (15)

SPC Barreda, Eduardo
PFC Behrens, Nicole
PFC Cardoza, Giovanni
SFC Coleman, Linda
SGT Faraon, Joe
SPC Fullen, Lonnie
SPC Garrett, Ramona
**SFC Hilliard, Ireshekia
SPC Holland, James
PFC Krouser, Michael
SPC Murphy, Michael
**CPT Pinzon, Cristian
SGT Wildeman, Brandon

SPC Wildeman, Christina M.
SPC Wilson, Danny

Headquarters, 175th Adjutant General Company (Postal) (Direct Support) (2)

SGT Clay, Elizabeth
1SG Gassaway, Robert

1st Platoon, 175th Adjutant General Company (Postal) (Direct Support) (1)

PFC King, Alonzo

2nd Platoon, 271st Adjutant General Company (Postal) (Direct Support) (1)

**SPC Elamar, Quinones

2nd Platoon, 342nd Adjutant General Company (Postal) (Direct Support) (5)

SSG Brashears, Linda
SGT Cline, Terrence
SGT Gomez, Hernan
SGT McNeil, Travis
SFC Williams, Tommie

Detachment 5, 376th Personnel Services Battalion (2)

SSG Holdaway, April
**SSG Miller, Joshua

2nd Platoon, 458th Adjutant General Company (Postal) (Direct Support) (1)

1LT Lewis, Edward

1st Platoon, 747th Adjutant General Company (Postal) (Direct Support) (3)

SSG Campbell, James
SPC Miech, Kenneth
SGT Mostad, Kyle

1st Platoon, 834th Adjutant General Company (Postal) (Direct Support) (4)

**SSG Clamens, Lillian L. *
SFC Hammond, Reginald
SGT Perez, Monica
SSG Soto, Yuliana

Detachment 4, 847th Personnel Services Battalion (5)

**SSG Edjou, Fabien
SGT Garcia, Moises
SGT Roberts, Edward
**SFC Tobiason, John J. *
SGT Warner, Ryan

**Purple Heart Recipient

*Killed in Action

NOTE: This listing is based on internal record keeping.

BRONCO STRONG'S GLOSSARY

- 1SG – First Sergeant
- 2LT – Second Lieutenant; the Entry Rank for Commissioned Officers
- ACU – Army Combat Uniform
- AG – Adjutant General's Corps
- AHA – Ammunition Holding Area
- AO – Area of Operations
- ARCENT – Army Central Command
- BAMC – Brooke Army Medical Center (San Antonio, TX)
- BC – Battalion Commander
- BDE CDR – Brigade Commander
- BG – Brigadier General is the First General Officer Rank
- BIAP – Baghdad International Airport
- CAB – Combat Action Badge
- CAC – Common Access Card
- CENTCOM – United States Central Command (aka USCENTCOM)
- CLT – Casualty Liaison Team
- COL – Colonel; the Sixth Commissioned Officer Rank
- COSCOM – Corps Support Command
- CPT – Captain; the Third Commissioned Officer Rank
- CSH – Combat Support Hospital

- CSM – Command Sergeant Major
- CSSB – Combat Sustainment Support Battalion
- CW2 – Chief Warrant Officer Two; the Second Warrant Officer Rank
- CW3 – Chief Warrant Officer Three; the Third Warrant Officer Rank
- CW4 – Chief Warrant Officer Four; the Fourth Warrant Officer Rank
- DCIPS – Defense Casualty Information Processing System
- DD93 – Next of Kin form
- DFAC – Dining Facility
- DOA – Dead on Arrival
- Enlisted Ranks – There are 13 Enlisted Ranks in the U.S. Army: Private, Private Second Class, Private First Class, Specialist, Corporal, Sergeant, Staff Sergeant, Sergeant First Class, Master Sergeant, First Sergeant, Sergeant Major, Command Sergeant Major, and Sergeant Major of the Army.
- EOC – Emergency Operations Center
- EPMD – Enlisted Personnel Management Directorate
- EPS – Essential Personnel Services
- FAST – Forward Area Support Team
- FOB – Forward Operating Base
- G1 – Division-level Staff Position focusing on Personnel
- G6 – Division-level Staff Position focusing on Communications
- HHD – Headquarters and Headquarters Detachment
- HQ – Headquarters

- HR – Human Resources
- IED – Improvised Explosive Device
- KIA – Killed in Action
- LNO – Liaison Officer
- LT – Lieutenant; the Appropriate Way to Address a Second Lieutenant and a First Lieutenant
- LTC – Lieutenant Colonel; the Fifth Commissioned Officer Rank
- MAJ – Major; the Fourth Commissioned Officer Rank
- MG – Major General; the Second General Officer Rank
- MNC-I – Multi-National Corps-Iraq
- MND-B – Multi-National Division-Baghdad
- MND-C – Multi-National Division-Center
- MND-CS – Multi-National Division-Center South
- MND-SE – Multi-National Division-Southeast
- MNF-I – Multi-National Force-Iraq
- MNF-W – Multi-National Force-West
- MP – Military Police
- MRE – Meals Ready to Eat
- MSG – Master Sergeant
- NCO – Non-Commissioned Officer
- NCOIC – Non-Commissioned Officer in Charge
- NOK – Next of Kin
- OIC – Officer in Charge

- OIF – Operation Iraqi Freedom; a Protracted Armed Conflict that began with the 20 March 2003 Invasion of Iraq by a United Stated-led Coalition.
- OIF II – The Second Rotation of Forces to Iraq
- OIF 3 – The Third Rotation of Forces to Iraq Announced in July 2004
- OIF 4 – The Fourth Rotation of Forces to Iraq Announced in December 2004
- OIF 06-08 – The Fifth Rotation of Forces to Iraq Announced in June 2006
- OIF 07-09 – The Sixth Rotation of Forces to Iraq Announced in 2007
- OIF 09-11 – The Seventh Rotation of Forces to Iraq
- PCS – Permanent Change of Station
- PFC – Private First Class
- PH – Purple Heart
- PSB – Personnel Services Battalion
- PSDR – Personnel Services Delivery Redesign
- PSS – Personnel Service Support
- PT – Physical Training
- Purple Finger – Semi-Permanent Ink or Dye that is Applied to the Forefinger (Usually) of Voters during Elections
- PVT – Private (different levels, also PV1 and PV2)
- PX – Post Exchange (retail store)
- R5 - Reception, Return To Duty (RTD), Rest & Recuperation (R&R), Replacement, and Redeployment

- R&R – Rest & Recuperation
- ROTC – Reserve Officer Training Corps
- RTD – Returned to Duty
- S1 – Battalion or Brigade-level Staff Position focusing on Personnel
- S2 – Battalion or Brigade-level Staff Position focusing on Intelligence
- S3 – Battalion or Brigade-level Staff Position focusing on Operations
- S4 – Battalion or Brigade-level Staff Position focusing on Supplies
- S6 – Battalion or Brigade-level Staff Position focusing on Communications
- SFC – Sergeant First Class; the Seventh Military Enlisted Rank
- SGLI – Service member's Group Life Insurance
- SGM – Sergeant Major is the Highest Military Enlisted Rank
- SGT – Sergeant; the Fifth Military Enlisted Rank
- SPC – Specialist; the Fourth Military Enlisted Rank
- SPO – Support Operations Section
- SSG – Staff Sergeant; the Sixth Military Enlisted Rank
- TMC – Troop Medical Clinic
- TOA – Transfer of Authority
- Troop Surge – A Troop Surge began on 10 January 2007 that increased the troop levels to more than 150,000 and ended 16 July 2008.
- T-Wall – Tall Concrete Walls Used for Force Protection

- USCENTCOM – United States Central Command (aka CENTCOM)
- VTC – Video Teleconference
- WO – Warrant Officer; the First Warrant Officer Rank
- XO – Executive Officer; the Second in Charge of the Battalion

Made in the USA
Monee, IL
04 November 2026

67378953R00148